The Synfuels Manual

The Natural Resources Defense Council is a nonprofit membership organization dedicated to protecting America's endangered natural resources and to improving the quality of the human environment. With offices in New York City, Washington, D.C., San Francisco, and Denver, and a full-time professional staff of lawyers, scientists and environmental specialists, NRDC combines legal action, scientific research, and citizen education in a highly effective environmental protection program. NRDC's major accomplishments have been in the areas of energy conservation, nuclear proliferation, toxic substances, air and water pollution, urban transportation, natural resources and conservation, and the international environment. NRDC has 45,000 members and is supported by tax-deductible contributions.

The Synfuels Manual

A Guide for Concerned Citizens

Edited by
Jonathan Lash
Laura B. King

Natural Resources Defense Council
New York, N.Y.

Library of Congress Catalog Card Number: 82-24538
International Standard Book Number: ISBN 0-9609358-0-0

Production and design: Peter Borrelli
Cover design: Stephen Jenkins
Typography: Word Management Corporation, Albany, N.Y.
Printing: Rumford National Graphics, Concord, N.H.

Natural Resources Defense Council
122 East 42nd Street
New York, N.Y. 10168

1725 I Street, N.W.
Washington, D.C. 20006

25 Kearny Street
San Francisco, California 94108

Contents

Acknowledgements

This book reflects the efforts of numerous, generous, and overworked individuals. More than a dozen members of the staff of the Natural Resources Defense Council contributed to it, including: Jim Banks, Peter Borrelli, David Doniger, Lisa Gollin, Joyce Goldstein, David Hawkins, Lawrie Mott, Wayne Naill, Bill Painter, Barbara Pratt, and Jackie Warren. In addition, Laura Ackerman, Dave Bradley, Norman Dean, James Evans, Nate Karch, and especially David Masselli and Cary Ridder made valuable, thoughtful, and important contributions, but are not responsible for any errors or misjudgments that may appear in the final product. Tom Keenan, a synfuel maven and a person of voracious curiosity and great intelligence spent countless hours digging out intriguing facts, writing readable prose, and goading everyone else to get the work done.

Our work on this book was supported by a grant from the Needmor Fund. We are grateful for their support.

Jonathan Lash
Laura B. King

1

Introduction

The alchemists strove but never managed to make lead into gold. Chemists and engineers, however, have discovered a remarkably practical alchemy. They can make dirty lumps of coal and oil shale into convenient liquid and gaseous fuels—synthetic fuels.

Nearly two centuries ago, gas made from coal illuminated London. Now some say synfuels, the energy alchemy of shale and coal, should be revived to provide fuel for the United States. The revival requires, they say, the tonic of federal subsidies. A multibillion dollar federal corporation has been created to administer those subsidies. Construction of several huge plants has begun with federal aid, and others are vying for billions of dollars in additional subsidies.

This book is for people whose lives and communities may be affected by the creation of a new synthetic fuels industry in the United States. We have endeavored to describe the potential effects of the construction and operation of synfuels plants on surrounding communities. Pollution, if not adequately controlled, may threaten the health of those living around a plant. The influx of construction workers, if not planned and paid for in advance, can turn farm communities into problem-plagued boomtowns.

The synfuels industry has a ragged history. It has boomed, and collapsed. It has provided fuel cleaner to burn than the coal from which it was made, but it also presented some of the earliest and harshest evidence of occupational cancer. Synfuels plants can be dirty and dangerous, but with adequate controls they can probably be built to be no worse than chemical plants or refineries. Since there are no commercial scale synfuels plants operating in the United States, we have too little experience and too little information to feel comfortable with assurances that synfuels plants will be safe and clean.

For those concerned about the effects of synfuels development, this book provides suggestions about ways to influence the process by which federal, state, and local governments review and authorize construction of synfuels plants. Citizens can have a significant impact. Citizen participation is especially important now as the federal government retreats from its responsibilities for environmental protection. The federal Environmental Protection Agency (EPA) has abruptly cancelled research on the health and environmental effects of synthetic fuels and abandoned efforts to develop guidelines for pollution control equipment to be installed on synthetic fuels plants. Since many of the hazardous substances that may come out of synfuels plants are not specifically regulated, the retreat of EPA is particularly troubling.

The Synthetic Fuels Corporation, the agency that administers federal synfuels subsidies, has taken the view that its job is fundamentally that of a banker, not a regulator. Thus, it has done little to assess or mitigate the potential effects on surrounding communities of the plants receiving federal aid.

Much depends upon whether local citizens decide to play a role. If they are organized and active, their impact can be significant. They can influence decisions about whether to build a plant, where to build it, what safety features are built in, what the sponsor does to train and employ local citizens, and how the sponsor helps to prepare and compensate local communities for the economic strains that accompany the development of massive new projects. But only active and informed citizens will be heard.

We have not endeavored to write an even-handed discussion of the pros and cons of synthetic fuels. This book describes the problems that may concern individual citizens living near proposed plants and suggests some ways to address those problems. Sponsors of synthetic fuels plants are quite capable of arguing the benefits that synthetic fuels development may bring.

However, we would note that to be concerned about the impacts is not the same as being "anti-synfuels." *We believe that the active involvement of well-informed citizens can only improve the quality of both government and private industry decisions.*

A Question of Policy

There may not be a synfuels industry anytime soon. The small experimental plants that are now in operation have been built with federal funds. The few commercial scale facilities under construction also have federal subsidies. Dozens of proposals for which the private sponsors received or sought federal planning grants have been shelved, as has the largest federally assisted project, the proposed 100,000-barrel-per-day Colony shale project that was under construction in western Colorado.

Why has the bubble burst so soon? There are two reasons. First, oil prices, which rose ten fold during the past decade, have declined slightly, as recession and conservation reduced demand for oil. Second, federal money, while still flowing to the synfuels industry, is not flowing so freely. The big energy, chemical, and engineering firms that proposed to build synfuels plants have concluded that they cannot produce synthetic fuels at a price to compete with crude oil and natural gas. The fate of the synfuels industry seems to be in the hands of the federal government.

The rest of this book provides information about the potential impact of synfuels plants if they are built, but first it makes sense to address briefly the underlying question: should federal subsidies be provided so that synfuels plants get built? We will begin with an account of how the present subsidies came to be enacted and how they work. We will then discuss what the subsidies may achieve. And finally we will examine some alternatives.

The federal government has supported research on synthetic fuels sporadically since the 1920s. During the 1970s, direct federal grants supported the construction and operation of a number of small experimental plants. By 1979, the Department of Energy budget for synfuels was about $1 billion. Several events took place in 1979 that led to a sudden expansion of the federal synfuels program. The fall of the Shah of Iran exacerbated Middle Eastern instability. Oil prices began to rise steeply. At the same time, the Carter Administration decided to phase out price controls on domestically produced crude oil and to seek a windfall profits tax to capture a portion of the resulting revenues. In the spring of 1979, spot shortages of gasoline led to gasoline lines and intense pressure on the federal government to do something, or at least appear to be doing something.

The House of Representatives swiftly passed synfuels subsidy legislation. Advocates for the program argued that a synfuels industry would assure the United States of adequate supplies of domestically produced liquid fuels, thus protecting our future and enhancing national security. Private industry, they said, could not be expected to make the huge investments necessary to build synfuels plants, when world oil prices were controlled by an unstable cartel.

Shortly afterwards, the Carter Administration leapt on the bandwagon with an awesome proposal to spend $88 billion by an independent federal corporation. The money was to come from the proceeds of the windfall profits tax.

Shortly afterwards, the Carter Administration leapt on the bandwagon with an awesome proposal to spend $88 billion through an independent federal corporation. The money was to come from the proceeds of the windfall profits tax.

The legislation passed, though only $20 of the $88 billion was immediately committed. Two years later, some of those who helped to shape the legislation observed that "it was symbolism and not substance that drove the synfuels legislation . . ."* Members of Congress "became convinced that the legislation could be sold to the public quickly . . ." Not only national security but old-fashioned, pork-barrel security banner, many legislators seized the chance to secure energy project commitments in their districts. Even conservation and solar energy enthusiasts in the Congress who were skeptical of synfuels seized upon the measure as a vehicle to gain support for bigger and better programs in their interest."**

The legislation that was passed, the Energy Security Act of 1980, established the U.S. Synthetic Fuels Corporation. The corporation is entirely federally funded. It is managed by a presidentially appointed board and its

---

***Energy Policymaking, Public Opinion and the Media: A Problem in Governance,* Summary of an Aspen Institute Workshop held September 11-13, 1981, p. 9.

***Ibid.* p.9-10.

principal function is to administer federal subsidies for synthetic fuels. However, it is exempt from most of the federal laws that require government agencies to operate in a fair and open way, and it is not subject to annual congressional budget scrutiny.

The Energy Security Act sets a goal of producing 2 million barrels a day of synfuels by 1992. It authorizes the Synthetic Fuels Corporation to provide loan guaranties, price guarantees, loans, and, in certain circumstances, to participate in joint ventures for the construction of synfuels plants. To hasten implementation of the program, Congress authorized the Department of Energy to administer the subsidies until the corporation was fully operational. The act also established much smaller programs to promote energy conservation, solar energy, biomass fuels, and geothermal energy.

In the fall of 1980, the Department of Energy issued several hundred million dollars of grants for industry applicants to develop synfuels proposals. With the election of Ronald Reagan, who opposed the synfuels program during his campaign as a wasteful giveaway, there was speculation that the program might be abolished before it was well launched. Synfuels funds were in fact reduced, but the president personally approved $3.6 billion in loan and price guaranties for two shale projects in Colorado and a synthetic natural gas project in North Dakota. No cuts have been made in the funds controlled by the Synthetic Fuels Corporation, and the corporation is negotiating with two projects preparing to make its first commitments.

The other programs established by the Energy Security Act have not fared so well. The Reagan Administration has sought to end virtually all federal conservation and solar energy programs, including research, information, and consumer assistance. The Administration has simply refused to carry out the conservation and solar portions of the Energy Security Act.

Could We Have Bought Better Insurance?

The synfuels program is essentially justified as an insurance policy against future oil shortages and price increases. The question is whether for the price we could purchase a better policy. To answer that question, we must first look at what the synfuels program will get us.

No one now believes that the program will produce anything close to the 2 million barrels per day by 1992 that is the objective of the Energy Security Act. If oil prices begin to climb again, production may reach half that level by 1992, and synfuels will provide an expensive 3% of our energy needs. Synfuels production at that level will not end our reliance on oil imports, which are now about 5 million barrels per day. The experience operating the plants would help in deciding whether and how to build more plants if the need arose, but would not get additional plants much more quickly. Synfuels plants will be built when their products are cheaper than the alternatives.

If oil prices do not rise steeply, the taxpayers will, in some future year, have to make good on up to $16 billion of loan and price guaranties. In a remarkable interview with *Synfuels Week*, the senior economist for the Synthetic Fuels Corporation said that the taxpayers will pay. The story continued:

> It would be to taxpayers' advantage, Ruff noted, if the SFC has to spend a lot of money on its first round of projects, because this would mean conventional fuels will have become plentiful and cheap. Ruff said he personally would like to see the Corporation fail; failure to bring about commercial synfuels production will mean only that the energy market does not need alternate fuel supplies.*

Those obligations do not appear in the federal budget now; some future Administration will have to meet the obligations incurred by the corporation. The immediate effect, however, is to propel synfuels projects into the credit market to compete with unsubsidized private borrowing.

As an insurance policy, the synfuels program does not offer much. Its coverage is extremely limited, its cost is high, and its benefits, in the event that a major oil shortage occurs, will be inadequate to cover the nation's needs.

Are other insurance policies available? The answer is yes. An energy conservation program on the same scale as the synthetic fuels program would achieve more and take effect sooner. In a study released in April 1982, the Oak Ridge National Laboratory concluded that the $1.8 billion spent on federal energy conservation programs since 1973 have already reduced U.S. consumption by five quadrillion Btus per year, the equivalent of 2.5 million barrels of oil per day.**

---

*"Failure at SFC projects could mean good news," *Synfuels Week*, 8 February 1982, 2-3.
**Hirst, Goeltz, and Trimble, *Energy Use From 1973-1980: The Role of Federal Conservation Programs*, Oak Ridge National Laboratory, 1982.

Thus, a program one-fiftieth the size of the synthetic fuels program has exceeded the goals that the Energy Security Act set for the Synthetic Fuels Corporation. Energy saved is just as good as energy produced in reducing U.S. dependence on foreign sources. Energy conservation has one huge advantage, however. Once the initial investment is made, it saves money instead of costing money. If, for example, we were to spend $1 billion weatherizing every school in the United States that is heated with oil or gas, those schools would reduce their consumption of oil and gas, probably by about a third. Not only would that lessen our need for fuel imports, it would mean those schools would have lower heating bills year after year. The billion-dollar investment would come back to local taxpayers many times over.

If we spent the same sum subsidizing construction of a synfuels plant, not only would it yield less energy, but when the plant began to operate we would have to pay for the energy it produced at the full market price.

Energy conservation measures do not cause pollution (wood stoves, which do cause pollution, are not conservation measures; they are simply a means to burn a renewable fuel). Conservation investments generate more jobs in a wider variety of fields than construction of synfuel plants. Conservation will benefit each of the millions of consumers and businesses that take steps to use energy more efficiently, whether oil prices rise or not. Synfuels plants are a far more risky proposition.

2

History

It may come as a surprise to discover that oil shale, coal gasification and even coal liquefaction are not new, untested technologies. Over the past two centuries, people in the Western world have tried to change plentiful and dirty solid fuels into useful and clean liquid or gaseous energy. More often than not, such efforts have been abandoned. Some of the failure may be attributed to the competition over the last century from cheap and abundant supplies of natural gas and petroleum. But just as often, the failure has stemmed from an inability to make complicated machinery work safely and effectively. Even during those brief episodes of success to which the modern synfuels industry often points—the large town gas industry of the 19th century and the German synfuels program during World War II—synfuels were expensive, inefficiently produced,and unsafe.

There is a tendency to believe that this time things will be different, that technology and vigilence will overcome past problems. And, indeed, our story suggests that this has often been the case. Things *are* different: hundreds will not die in Colorado shale plants, as they did in earlier in Scotland. Yet, the similarities remain powerfully evident, and our chief reason for beginning with the history of synfuels is to catch those similarities, trace the patterns of behavior of those who pursue the dream of turning rocks into energy. Whether in 1780 or 1980, national security and the desire to use indigenous resources are advanced as justifications for major projects. Whether it is the first Washington Gas Company in the mid-1800s, German industrialists, or today's oil giants, government subsidies are sought. Whether it is sooty Manchester a century ago, the Kosovo, Yugoslavia synfuels plant today or Western Colorado tomorrow, occupational health and environmental degradation are major issues.

We start this book with a thorough examination of the history of synfuels, then, for more than just academic reasons. The history is often fascinating, and it reminds us of an unfortunate past which has been changed and too quickly forgotten. It provides an easy introduction to some basic concepts ahout how synfuels work and do not work. The history also gives a good idea of the problems which still face the modern industry: pollution, worker health and safety, expenses and efficiency.

OIL SHALE

The oil is there, and we know how to get it out, and all you have to do is heat the rock, and the oil will come pouring out of the mountain.

—Armand Hammer

The oil, it seems, has almost always been there, and Armand Hammer's all-you-have-to-do reflects the history of people's attempts to deal with that fact. But the rhetoric of discovery may be somewhat misleading, for shale oil has been in shale rock for a long time, and it has proved somewhat more resistant to extraction than might be expected. In fact, today's attempts to heat the rock and get the oil have a history behind them that belies the notion that shale is a wonder fuel of the future.

The glossy high points of shale's past will not be emphasized here. We know that shale flew the first jet plane on its first jet flight.[1] We know that shale has always had friends in high places: one of the seven sisters, British Petroleum, was even born as a shale oil monopoly.[2] And we know that patents for shale oil production process identical in most respects to those employed today have been around since 1694, predating crude petroleum by a century and a half. But all this cannot be taken to indicate that shale has a solid 300-year history of success. On the contrary, even after all those years, after all that heating and pouring, each new shale venture still feels like the first one. And as we will try to show, it is not without reason that today's shale industry has forgotten most of its history, and remembers what it does only selectively.

But the history of shale oil does stretch back for some time—more than 500 years, by some accounts, to the middle of the 14th century. What was called *petroleum* (literally, rock oil) first emerged not as crude from the ground, as we know it, but as the product of heating rocks—shale rocks—until they broke down into ash and oil. The rocks were Austrian and Swiss shales, and after the raw stuff had been produced, it was processed to get *icthyol* or *fish oil,* because the rocks contained fossils of fish. It was purported to have medicinal properties.[3] Likewise, shale oil was used to make alum in the 1600s in Sweden, a date some people (mostly Swedes) claim as that of shale oil's birth.[4] At the end of the sixteenth century, Teutonic medicine men used shale oil to heal everything from sword wounds to cuts and scrapes.[5] Shale oil, today associated with environmental destruction and the threat of cancer, came into this world as an agent of health, a salve for wounds.

Beginnings

In North America, the story goes, shale's usefulness as fuel was discovered accidentally. The Ute Indians, for instance, discovered the energy in shale when they watched outcroppings burst into flames after being hit by lightning. They called it "the rock that burns," and they are supposed to have astonished innocent white visitors by lighting campfires which contained not wood, but rocks.[6]

White pioneers, indisposed to learn things from savages, came upon the remarkable fact of shale's combustive powers in a more difficult, and perhaps painful, manner. Three shale stories circulate which make this point, with some cleverness. One has it that a settler of the semi-arid Western shale regions in the 1800s chose to use the good solid rock which he found everywhere around him in constructing a sturdy pioneer cabin. When he had finished his work, he inaugurated his home with a roaring fire in its new stone hearth, only to watch it too catch fire and burn, quickly incinerating the entire cabin.[7] Likewise, the March, 1874, issue of *Scientific American,* reported that:

> about 800 miles west of Omaha the Union Pacific Railroad crosses the Green River, and the approach to the river is for a considerable distance through a cutting made in rock. During the construction of the road some workmen piled together a few pieces of the excavated rock as protection for a dinner fire and soon observed that the stone itself ignited. The place thereafter became know as Burning Stone Cut.

Always ingenious, and not to be outfoxed by a dumb rock,

> the general superintendent of the railroad, Mr. T.E. Sickels, has caused analyses and experiments to be made with this substance which proves to be a shale rock rich in mineral oils. The oil can be produced in abundant quantities, say 35 gallons to the ton of rock. The oil thus obtained is of excellent quality.

And even America's most famous explorers ran into oil shale for themselves. As Tom Bethell explains, "Lewis and Clark did the early research in the field. In 1805, they lit a campfire next to an oil shale cliff. The cliff began to smolder."[8]

The troubles began with the discovery of shale oil as a fuel, rather than simply a medicinal balm, which we can trace to 1694, when the British Crown issued a patent to one Martin Eele and colleagues, who had come forward with a way of removing and producing oil from, they reported, what seemed to be "a sort of stone."[9]

> "Being," as it reported, "willing to cherish and encourage all laudable endeavours and designes of all such persons as have by theire industry found out useful and profitable art, misteryes, and invencons," the British Crown granted Patent No. 330 to our friend Mr. Eele, who, along with his associates Hancock and Portlock, extract and make great quantityes of pitch, tar, and oyle out of a sort of stones," of which there is plenty within our Dominions . . ."[10]

In spite of this early technological advance, though, commercial mining and processing of oil shale did not get started until the middle of the nineteenth

---

Not all shales yield the same amount of oil or kerogen when heated. To compare the richness of shales, a test called the Fischer Assay is commonly used. In the Fischer Assay, a small amount of shale is heated in a closed vessel. Fischer Assay results are given in terms of gallons (not barrels) per ton. The so-called rich shales of Colorado generally have a Fischer Assay of 25-35 gallons, which is less than one barrel per ton. Eastern U.S. shales generally yield less than 15 gallons. Some foreign shales are much richer, although not as thick as the Colorado shales. For example, some shales in Australia yield up to 180 gallons per ton, South African shales average 55 gallons and Estonian shales average better than 50 gallons per ton. It is important to remember that the Fischer Assay is just a test; some retorting processes may actually yield a higher output than the Fischer Assay.

century. Mining began in France around 1838 or 1840, and by the 1850s a sizeable shale oil industry had also developed in the eastern United States. Here, shale oil had so effectively replaced both wood as the industrial fuel of the fledgling American economy and whale oil as America's choice lamp fuel that by 1859 at least 50 commercial facilities operated along the Atlantic seaboard.[11] In 1859, "Colonel" Edwin Drake also drilled the first commercial oil well in Titusville, Pennsylvania. Within a matter of years, the Eastern oil shale industry collapsed. Very little, it seems, is known about this episode, and as we will see later, the relation of the industry to the emerging use of coal on the Eastern seaboard remains unclear.

Shale Around the World

Just as the U.S. industry was crumbling, a fairly long-lived shale industry was being born in Scotland. The first retorting plant was built in 1859, but within a few years more than 140 companies and individuals had projects going. As a fuel-oil industry, though, its life was short. When faced with competition from cheaper petroleum, the Scottish shale industry survived by producing byproducts such as waxes, ammonia, and building materials.

At its height in 1913, six oil shale companies mixed 3.3 million tons and produced about 4,400 barrels per day.[12] Shortly thereafter, Sir William Fraser consolidated these companies into Scottish Oil, Ltd. which later became a subsidiary of Anglo-Iranian Oil which in turn became British Petroleum.

The Scottish industry encountered a steady decline as the 20th century wore on. It was given special tax breaks, but production declined to 1.4 million tons in 1947, and the last plant was closed in 1962.[13]

The Scottish shale industry subsisted on relatively low-grade, 22-gallon-per-ton shale, but in Australia there were fabulously rich deposits with yields as high as 180 gallons per ton. Production began in 1862 and grew intermittently until 1917 when the Australian government began to subsidize the industry. In 1938, the National Oil company built an 11,000-barrel-per-day plant near Glen Davis, New South Wales.[14] The plant never operated at that figure (total *national* production peaked at 3,000 BPD in 1947) and was abandoned in 1952 after the richest shales were exhausted. At its height, the industry provided about 3% of Australia's gasoline.

Three other countries still have existing shale industries. In Brazil, production from the Irati shales south of Sao Paulo has taken place since 1862. Production has always been insignificant. In 1970, Petrobas, the national oil company, completed construction of a 2,200-ton-per-day demonstration plant using the Petrosix retorting process.[15] Petrobas has plans for a larger plant, on the scale of proposed U.S. commercial projects, but has not formally committed itself to construction.

In the People's Republic of China, a large shale deposit is located at Funshun in Manchuria. Though the shale is relatively lowgrade (about 15 gallons per ton), it overlies one of the richest coal seams in the world. When the Japanese invaded Manchuria in 1926, they began to develop the shale, which had previously been discarded as overburden from the coal mine. About 1.3 million barrels of oil were recovered during the period of Japanese domination.[16]

After the war, the Chinese modernized the Funshun operation and built a second complex. During the Korean War, production reached approximately 50,000 barrels per day; byproducts including fertilizer from ammonia and cement from spent shale were also produced.[17] It is estimated that current production from a series of retorts remains at about that level.

The largest existing shale industry appears to be located in the Soviet Union, almost exclusively centered in Estonia. Estonian shale, known as kukersite, averages 50 gallons per ton with reasonably large reserves. Production began in 1925 when the State Oil Shale Industry built its first retort and by 1938, with annual production of 800,000 tons per year, the industry was the second largest in the world after Scotland.[18] The Germans occupied Estonia in 1941, but not before retreating Russian troops disabled a major portion of the industry. The Germans quickly rebuilt much of it and were able to garner some production before losing it back again in 1944.

The Estonian shale industry has grown rather steadily since then, but almost three-fourths of its output is burned directly in specially designed electric power plants where it provides about 90% of Estonia's electric needs.[19] In 1980, Estonia produced about 31 million tons of shale. Only about 20% of Estonia's shale is changed into liquid fuel, amounting to a production rate of around 25,000 barrels a day.[20]

Looking over the experience in the six foreign countries where shale industries have been tried, it is clear that no one else in the world is any closer than the United States to developing a working oil shale industry. No one has built even a single plant that produces 10,000 barrels per day; a plant that would be a module in any modern U.S. facility. Nor has any industry successfully surmounted the economic difficulties inherent in retorting and producing shale. It is noteworthy that the Soviet industry primarily uses shale for direct burning,and the key to the Chinese industry was the availability of mined shale as an essentially "free" byproduct of the world's largest coal strip mine.

Oil Shale in the Western United States

At the beginning of the 20th century, one of the periodic oil shortages and accompanying panics that have swept America since the 1880s was touched off when supplies of oil could not meet demand and imports increased. The U.S. Geologic Survey announced that there was only a nine-year reserve of domestic oil,and the outlook for new discoveries was bleak. The agency also announced that large quantities of shale were available in the Green River Formation of Colorado and Utah.[21] The first shale boom was on.

At that time, shale was classified as a "hard-rock" mineral under the Mining Law of 1872. By the terms of that law, prospectors who located a "valuable mineral" on public lands (lands owned by the U.S. Government) could stake a claim. When that claim was patented—on making the proper legal showing that the requirements of the act had been satisfied—the land on which the claim was located would become property of the prospector. Spurred by the U.S.G.S. announcement, prospectors headed for Western Colorado and staked more than 30,000 claims before 1920, when the law was changed to make oil shale available from federal lands only by lease.[22] The change in the law—made by the Mineral Leasing Act of 1920—did not affect the legitimacy of the claims that had been filed before 1920, and hundreds of thousands of acres of valuable shale land were transferred to private ownership.[23]

Some 200 shale companies sprang up in response to this oil shortage. After all, oil was expensive and land was cheap.

"Suddenly the hills were dotted with oil shale plants. There was no best way to extract the oil. Each operator followed his own instincts. But the boom soon went bust. It wasn't easy to turn a profit, and most companies went bankrupt."[24]

Of the 200 companies which formed, as least 25 got as far as building pilot plants.[26] Very little oil was actually produced, in spite of such high interest. The old boom ended abruptly with the discovery of large oilfields in east Texas. Oil prices dropped to a few cents per barrel, and interest in oil shale development essentially disappeared."[27] Local residents know the pattern: "Like, say, the oil shale's been in here for fifty, sixty years, and everyone that's come in has said, well, this is going to be a commercial thing in three or four years or five years or whatever, and it's never happened, so most of [us] don't believe it; [we] still don't believe it."[28]

But not all of the shale entrepreneurs were small capitalists. Cities Service, Standard Oil of California (Chevron) and Texaco began acquiring large blocks of shale properties in 1918,and they were joined by Union Oil in 1920. Union ended up with the largest block, over 33,000 acres. Except for some small efforts by Chevron in the 1920s, none of these oil giants actually tried to produce shale oil or do research and development during the first shale boom. They just gathered up shale-rich blocks of land and left the risk taking to the small companies and the U.S. Government.

Between 1925 and 1929,the U.S. Bureau of Mines operated a small research facility in Colorado, but it attracted little interest until 1944 when Congress, concerned about developing oil supplies for wartime use, passed the Synthetic Liquid Fuels Act authorizing the construction and operation of demonstration plants to produce synthetic fuels from coal and oil shale. Congress authorized a total of $87 million for the program, of which about $18 million was spent on the Anvil Points oil shale project near Rifle, Colorado.

Bob Prather, longtime resident of the shale town of Debeque, Colorado, remembers the 1920s boom:

... an old retort from the 20s ... basically worked like a still; they'd put the oil shale in some kind of vessel, heated it up, set it on fire, and just cooked the oil out of it. It was mined from up on Mount Logan three or four miles away, and it came off the cliffs up there with a tramway. Then they hauled it in here with teams and wagons. There was very little use for the oil and they didn't make very much of it. The cars, there were probably half a dozen in this community at that time, and it was used for veterinary purposes, doctoring cuts on horses, their hooves, that type of thing ...

In the early days, there were some kind of, I guess you would call them fly-by-night oil shale schemes—where they-it was a promotion. You know, they promoted an oil shale plant and everything, then they got stockholders to come in and they took their money and ran. Some of that happened; some of these old oil shale companies that you hear about, they actually never did anything except sold their stock and left.[25]

The law put the Bureau of Mines back in the shale business. They began by building two 40-ton N-T-U retorts at Anvil Points in 1946. The plants operated from May, 1947 to June, 1951 and produced a total of 20,300 barrels in abut 13,500 hours of operation.[29]

Twenty years later, in 1972, the government leased the Anvil Points site to a consortium of 17 companies, called the Paraho Oil Shale Project, to test a new gas combustion process owned by Development Engineers, Inc. According to the DOE environmental impact statement on the project, 10,000 barrels of shale oil were produced in 1975, refined into gasoline and jet fuel, and tested in military vehicles and ore freighters on the Great Lakes. Between 1976 and 1978, the project generated about another 100,000 barrels, and the semiworks plant closed in early 1978 (the smaller pilot plant has been used intermittently for tests since then). Still, even the larger semiworks retort was less than one-twentieth the size of the commercial size plants that most shale entrepreneurs seek to build. Test runs were intermittent; Paraho's longest run, 105-days, set the record. It would have been a disaster in a commercial plant: one-third of a year at one-tenth of retort capacity.

Efforts at commercialization have traditionally stopped just before the point where major financial commitments were required. In the early 1960s, Union attempted to sign a long-term fuel oil contract with a Southern California utility. The deal fell through and although Union has announced two major process improvements with great fanfare in recent years, its last major experiment ended 24 years ago. TOSCO ended a seemingly successful semiworks test in 1972. Shortly thereafter, it announced a plan to construct a commercial plant. But 10 years later the commercial plant is still off in the future and no test has ever been made at the 10,000-barrel-per-day module size.

Bolstered by recent infusions of federal supports Union now says it is moving forward to construct large plants. Other would-be shale developers have lined up at the door of the Synthetic Fuels Corporation for further federal assistance.

GAS FROM COAL

Coal gas provided the first gas-heat and gas-light across the world, preceding natural gas in the same way that shale oil paved the way for petroleum as an industrial fuel. Yet, coal gas survived much longer, and played a greater role than shale oil in the development and degradation of the Western world. Shale oil was replaced by crude soon after it came into use as a fuel, but coal products retained the advantages of wide availability and easy transportation well into this century. Only when pipelines were built could natural gas move long distances and displace coal from widespread small-scale, local use. For a long time, when one said "gas" in Europe and America, one meant coal gas.

Carbonization in 19th Century England

Coal gas was born in the back-room laboratory of a Yorkshire minister and part-time alchemist, John Clayton. Sometime in the late seventeenth century, Clayton attempted to replicate at home the widely reported phenomenon of naturally-produced gas in coal seams, notable both for its dangerously explosive tendencies and its illuminating powers when captured in crude mine lighting systems. In his alchemist's retort, a closed vessel, the reverend heated up some local coal and had his suspicions confirmed when a luminous and inflammable gas was produced.[1] Clayton described his discovery as a "spirit which issued out and caught fire at the flame of candle."[2] This "spirit of coals" Clayton collected, and, according to one historian, "amused his friends by lighting the gas as it escaped."[3]

Once Clayton's findings were reported to the Royal Society some 50 years later in 1739, the rush to gas was on. The process was simple, and remains to this day the basis of coal gasification technology; burn coal in the absence of air, and you get a gas product. Interest was spurred by the recognition that coal gas was commonly generated as an everyday byproduct in the production of coke (for use in iron-smelting), which had emerged as a major industry in eighteenth-century England. If these waste gases could be captured and moved through a pipe to a definite place (like a burner), instead of being casually discarded to the atmosphere, there was a profit to be made. Inventors scrambled to find a way to capture the waste gas and use it for heating and lighting.

Efficiency was not the only motivating factor. Even then, pollution was a problem as England was fueled by coal, which brought with it dirt and death. As early as 1273, Edward I had banned the burning of sea coal in London, in one of history's earliest attempts to control pollution. In 1616, London's chronic smog had been traced to coal, and then linked with respiratory diseases. In " Smoake of London," John Evelyn had written of the "hellish and dismall cloud of 'sea coale' impure and thick mist accompanied with a fuligious and filthy vapor . . . corrupting the lungs."[4] Gas, on the other hand, seemed a clean fuel, and England needed a clean way to use coal.

By the middle of the 18th century, then, pushed by the problems of pollution and the scientific excitement of coal technology—problems and excitements which have yet to disappear from discussions of coal gasification—schemes for gasifying coal were everywhere. The real possibilities for commercial application of coal gas seem to have come to light about 1782, "when several explosions took place in the tar ovens" of the Earl of Dundonald's experimental British Tar Company. The discovery was, according to historians, "put to good use in the illumination of several rooms in Dundonald's house, Culross Abbey, [although] the company . . . was never financially successful and his efforts came to nothing."[5] He did, however, light the Abbey with coal gas in 1787.[6] Commercialization finally came with the more advanced work of one William Murdoch, who began with a small distillation operation of unique design: "he first produced gas by heating coal in a teapot borrowed from mother."[7] Further demonstration came "in an experimental iron retort," which he had installed in the back yard of his house in 1792, and from which he "conducted the resulting gas by a pipe to a burner fixed near the ceiling over a table in one of the rooms in his house; . . . [thus] the use of the gas as an indoor illuminant was demonstrated."[8] The milestones came quickly after this breakthrough, because Murdoch realized that coal gas of this sort could be centrally produced and piped short distances to points of use, where it provided better light than oil or candles. Murdock set up a small gas-making operation at a nearby ironworks in 1795-6, where gas was burned to give "a strong and beautiful light, which continued burning a considerable time."[9] Murdoch was hired by James Watt (the inventor of the steam engine) in 1798-9 to continue the experiment in the Soho Works (foundries) in Birmingham of Messrs. Boulton and Watt.

Regular gas generation and lighting followed in 1803 at the Soho factory, and were extended to Manchester cotton mills in another year. The refinements were coming fast now, it seems: "in these Manchester installations some purification of the gas was attempted, for it is recorded that the gas made therefrom did not have the 'Soho stink'."[10]

Though awarded the Rumford Medal of the Royal Society in 1808 for his gasification work, Murdoch was displaced in commercializing his invention by a "remarkable showman from Moravia, F.A. Winzer who, on reaching England, wisely anglicized his name to Winsor."[11] Winsor developed the gas grid, featuring centralized production and a large-scale distribution network. "The unadorned aim of this flamboyant character was to sell gas-lighting as widely as possible" says Colin Russell, and other historians echo the description. [12] According to Barash and Gooderham, Winsor and his colleagues were "enthusiasts and great publicists." Publicists perhaps, but not poets, as a Winsor attempt should illustrate.[13]

Must Britains be condemned for ever to wallow
In filthy soot, noxious smoke, train oil and tallow
And other poisonous fumes ever to swallow
For with sparky soots, snuffs and vapours men have constant strife
Those who are not burnt to death are smothered during life.

In order "to sell gas-lighting as widely as possible," Winsor held lectures and public displays, produced publicity pamphlets, and in June, 1807, even staged a lavish exhibition of gas-light (piped from his Pall Mall home) at the Carlton House Gardens, the Prince of Wales' summer home. That same year, he requested a Parliamentary charter for his own gas company; because, then as now, the sums involved were so large as "to justify a claim for a limitation of liability which, in those days, required a private Act of Parliament."[14] It took three years, and some Parliamentary wrangling, but Winsor's lobbying and public demonstrations finally paid off with a gas company charter in 1810 (backed by 200,000 English pounds capital, down from the first try of an even million pounds) "to provide streets, squares, and houses with gaseous lights by means of conducting tubes underground from distant furnaces on the same principle as houses are now supplied with water."[15] Winsor's London and Westminster Gas Light and Coke Company finally came into being in April, 1812, and by New Year's Eve of 1813 he had lighted London Bridge with coal gas. By 1830, 200 gasworks, or retort houses, were spread across England, providing light for churches and colleges, factories and mills, streets and homes.[16]

"Lighting London with Smoke" . . . and Cancer

By the 1840s, coal gas use had been established throughout England, but not without more government assistance. Besides supporting the finances of the first gas company, the Crown itself introduced coal gas in some localities, and a multitude of favorable governmental reports could not have hurt the industry's chances. Of course, coal gas still stirred up great commotion. Novelist Sir Walter Scott is said to have laughed about "lighting London with smoke," and English scientist William Hyde Wollaston said that Winsor "might as well try to light London with a slice of the moon." A Scottish poet penned this verse on the same subject:[17]

> *We thankful are that sun and moon*
> *Were placed so very high,*
> *That no tenpestuous hand might reach,*
> *To tear them from the sky.*
> *Were it not so, we soon should find*
> *That some reforming ass*
> *Would straight propose to snuff them out*
> *And light the world with gas.*

The reaction, though, was not simply blind resistance to change. "Particular anxiety was felt over the safety of large volumes of inflammable gas stored in gasometers, as they were called. . . . [And] there were indeed several explosions."[18] The early experiences with coal seam explosions, as well as the Earl of Dundonald's accidental discovery of gasified coal's explosive properties, were to repeat themselves all too often, not only in the retort houses. London's streets were also said to be unsafe both for lamplighters and pedestrians. A satirical cartoon captioned, "A London Nuisance: One of the Advantages of Gas Over Oil," appeared in a London paper in 1822 showing passers-by injured in an explosion in a chemist's shop window.

More recently it has been said that "certainly the nineteenth century saw plenty of unnecessary deaths, both from explosions, and from coal gas poisoning."[19]

This latter hazard resulted from the production of highly toxic carbon monoxide (CO) and hydrogen sulfide (H_2S), along with cyanogen, in uncontained and often open gas retorts. The gases were thus seriously contaminated; this posed a serious problem, since these chemicals, "rendered them both evil smelling ("the Soho stink") and poisonous."[20] When gas engineers attacked that problem by introducing a purification step, another serious problem arose. The gas was passed through a filter of lime; this "lime-washing" successfully removed the smelly sulfur compounds, but left the engineers with problems of the disposal of tremendous quantities of dangerous "spent lime."[21] So that method was abandoned, and a bath of iron oxide was tried. Again, the gas came out clean, but tons of "spent oxide" remained. [22] Significant pollution from this "smokeless fuel" also occurred until large-volume gas storage tanks were invented—because most gas works preferred to discharge any excess gas to the atmosphere during periods of low off-take, i.e. daylight hours. And once the discharge problem was solved with the invention of water-sealed gas holders, "there was the further complication of disposal of the contaminated water."[23]

It is no wonder, then, that two British historians described "the familiar look and smell of a gas works" in this way: "a dirty, hodge-podge of large, cumbersome, smelly, tarry plants."[24]

Yet, these public health and environmental problems—as serious as explosions, smells, poisons, and hazardous wastes were in the nineteenth century—were dwarfed by an even larger risk. The coal carbonization process yielded for each ton of coal about 15,000 cubic feet of low-Btu gas (commonly called "town gas"), dangerous quantities of CO and H_2S, and a coal tar rich in carcinogenic polyaromatic hydrocarbons (PAHs), other hydrocarbons, and phenols.[25]

The greatest known risk in gas retort houses seems to have come from the unprocessed tar, along with the other uncontained by-products of coal's partial combustion. The result was cancer. It took more than 100 years, but in 1931 the evidence of occupational hazards started rolling in, connecting coal gas plants with a high incidence of all sorts of cancer. The first reports came from S.A. Henry and N.M. and E.L. Kennaway, who found that British gas workers in the heyday of British coal gas during the last quarter of the nineteenth century, were "specially liable to develop cancer of the skin and bladder."[26] And again, in 1936 and 1947, the Kennaways, following seven groups of gas workers in England and Wales, discovered that "the number of deaths attributable to cancer of the lung was greater in each of the seven groups than the number expected from the experience of the total and the excess varied from 79% to 184%."[27]

Later in 1952, Richard Doll reported following London gas workers whose "employment went back to before the first World War and in many cases to the end of the nineteenth century."[28] He found dramatic documentation of deplorable conditions in the nineteenth century plants: "in the group studied, the mortality from cancer of the lung was 181% of the mortality which . . . would have been suffered by a comparable group of representative Londoners," and "the excess mortality from lung cancer occurred principally among employees engaged on the production of gas and on the treatment of waste products [197% of the expected figure]."[29]

Gas continued to be produced by carbonization or distillation, in essentially uncontained ovens (retorts), well into the twentieth century. The entire history of coal gas as practiced in England, from the "Soho stink" and explosions and poisons of 1798 onward into this century, has been one of significant risk. Although the early reports of skin cancer showed a danger so great that "preventative measures were undertaken and . . . these conditions have now [1965] been practically eliminated from the industry,"[31] other sorts of cancer hazards remained well into this century.

Two British scientific teams have performed large-scale mortality studies on workers in the twentieth century British coal gas industry (generally men who entered the coal gas industry in the 1920s), comparing death rates of workers with high exposure to gasification processes with national (England and Wales) rates. The first study, which looked at almost 11,500 regular workers for eight years, had three important conclusions:[32]

Coal tar, a by-product of all kinds of coal conversion processes, has a lengthy history of association with cancer, experimental as well as occupational. Coal tar was the first material shown to induce tumors in experimental animals and as early as 1875 skin and scrotal cancers were reported among the unfortunate German workers who handled the tar . . . findings confirmed by further study in 1908 and in report after report ever since.[30]

Of course, cancer has been associated with the carbonization of coal ever since 1775, when the infamous Sir Percival Pott published *Cancer Scroti* (subtitled, "In Chirurgical Observations Relative to the Cataract, the Polypus of the Nose, the Cancer of the Scrotum, the Different Kinds of Ruptures, and the Mortifications of the Toes and Feet"), which documented a veritable epidemic of scrotal cancer among unwashed London chimney sweeps.

1. the death rate from lung cancer was 69% higher than the national rate;
2. the bronchitis death rate ran 126% above the national rate;
3. although the actual number of deaths was smaller, death from bladder cancer, scrotal cancer, and pneumoconiosis was again more common than nationally.

Lung cancer and bronchitis deaths were also found to be much higher than the corresponding rates from the regions where the gas works studied were located, with excesses reaching 74% and 144% respectively.[33] The study was controlled for smoking as well as for the amount of exposure, leading to the conclusion that, for gas-workers, "exposure to the products of coal carbonization produced increased death rates from lung cancer and bronchitis, and that these rates increased with intensity of exposure."[34]

The follow-up study seven years later, compared death rates for about 3,000 of the original group plus another 4,500 from additional coal gas plants to national rates. Again, large excesses in death rates among heavily-exposed workers were found;[35]

> . . . the consolidated results for the whole 12 years of the investigation [show that] . . . the mortality rates from both lung cancer and bronchitis are highly significantly in excess of the national rates, [and] . . . the death rates from bladder cancer and from cancer of the skin and scrotum are also significantly higher than the national rates.

The new data provided "adequate confirmation that exposure to the products of coal carbonization can give rise to cancer of the lung,"[36] and confirmed the earlier findings that "work in a retort house involved a hazard of developing cancer of the bladder."[37] Doll also noted that scrotal and skin cancer had largely been controlled by industrial hygiene programs; the deaths from those causes dated back to "a time when precautions to prevent exposure were less effective than they have been since the second world war."[38] But industrial hygiene, Doll reported, only works when

---

Whether or not the work environment has been cleaned up of late, outdoor pollution continues. As recently as June, 1981, England's *New Scientist* reported that "Britain's smokiest grimiest—and probably, smelliest—factory," a South Wales plant which makes coal-derived "smokeless fuels" by a technology quite similar to gasification, remained as the country's No. 1 polluter despite attempts to close it down for many years, and 15 years of increasing pollution.

followed strictly; the excess of bronchitis was "at least in part attributable to [the] effects of the extremely unfavorable conditions of work in retort houses during the war years when a blackout was imposed."[39] As then-administrator of EPA Douglas Costle told the Synthetic Fuels Corporation in 1980, these "studies of workers exposed to poorly controlled coal gasification plants . . . reveal how potentially dangerous [they] could be."[40]

In spite of the hazards and pollution from coal gasification, though, the industry flourished for a long time. Colin Russell observed that "this form of illumination dominated the nineteenth century." Yet, he adds, "as the century grew to a close there came the new challenge of electric lighting,"[42] and later still came the discovery and rapid development of natural gas. In England, the gas industry responded first to electricity by inventing an incandescent gas mantle, a sort of gas light-bulb, advertised to have "all the advantages of Electric light and none of its drawbacks."[43] This device, introduced in 1890, "in fact enabled [coal] gas-lighting to continue side by side with electricity for several more decades."[44] But ultimately the town gas plant succumbed to natural gas—plentiful, cheap, high in energy, and easily transported by pipeline.

Europe and America

Paralleling the enormous and immediate spread of town and producer gas from coal in England was its expansion on the European continent and in the newly liberated American colonies. Very little information is available about the European gasification experience. Apparently, at about the same time that Reverend Clayton was amusing his friends "around 1681, Professor Johann Becker of Munich, Germany, discovered that combusti-

---

The factory belches black smoke, dust, smuts, fine droplets of tar, and heavy hydrogen sulphide fumes into the surrounding communities in the valley. An analysis . . . found that dust and grit from the plant was four times worse there than similar pollution anywhere else in the country. "We get the pollution here so people in London and Birmingham don't have it," says Phillip Williams, who lives just a stone's throw from the plant. . . . The local environmental health officer, Maldwyn Jones, says the factory is like "Dante's Inferno, Mark Two."

Phillip Williams sums up the local people's frustration. "The smell is like boiling tar being made up in the road. When the wind comes this way we can't sit out in the garden; we can't leave washing out. I painted the front of my house two years ago. Now it's black and the paint is being eaten away by tar.[41]

ble gases resulted from heating coal in the absence of air." As in England, coal provided the first widespread and easily distributed source of energy for heating, lighting, and cooking, and at about the same time as well.

Technologies for carbonizing coal into gas moved across the Atlantic almost as soon as they were commercially introduced in England.

> In the United States, Benjamin Henfrey demonstrated the use of gas for lighting in Baltimore in 1802 and again in the following year in Richmond, Virginia. There were other demonstrations, but David Melville of Newport, Rhose Island, seems to have been the first American to have gone beyond the demonstration stage when he lit his own house in 1807 and illuminated several cotton mills between 1813 and 1817.[45]

Within a short time the process had spread across the state and beyond. Baltimore was the first city in the country to get gas lights, after painter Rembrandt Peale and four other prominent citizens formed the Gas Light Company of Baltimore in June, 1816. Baltimore switched from pine tar to coal as the source of its gas in 1822, "with much more satisfactory results."[46] The same year Boston went the coal gas route. New York City followed in 1825, and "by 1828 New York's famous Broadway was lit by gas lamps."[47] Pittsburgh coal, said to be "the best in the world for gas light,"[48] seems to have provided much of the initial fuel (just as petroleum from another Pennsylvania town, Titusville, would later inaugurate the petroleum era). And it was shipped east on a special canal to light the Eastern seaboard, and later even supplied Chicago, Detroit, Milwaukee, as well as cities in Ontario, with coal for gas.

> In 1834, the city of Philadelphia decided to build its own gas works and in so doing became the first municipally owned gas company in the United States. Some other cities where manufactured gas companies were established around this period were Louisville and New Orleans in 1832, Pittsburgh in 1836, Washington in 1848, Brooklyn in 1849, Chicago in 1850, and Atlanta in 1855. The first issue of *The American Gas Light Journal,* published on 1 July 1859, listed 183 gas light companies in the United States: by 1870 a lighting service based on manufactured gas had been established in 46 American cities. . . . A census published by the *Gas Light Journal* on 1 March 1886, discloses that the number of gas light companies had grown to 971, and that a total of 186,901 public lights were now lit by gas. . . . The majority of the companies listed were manufacturing and selling gas made from coal.[49]

By the middle of the century, as America's mines moved west, local coal took over and cities and towns in Alabama, Kansas, Utah, and many other states got their light from their own coal.[50]

Of course, some had already made their fortunes in coal gas: "already by the 1860s . . . Thomas Mellon was a financial power in Pittsburgh, with interests in coal and gasworks."[51] Coal gas was expensive: "manufactured gas sold in 1860 for about $8/MMBtu," in 1860 dollars, while gas today goes for around $3/MMBtu (1981 dollars).[52] New York City gas cost $10 per thousand cubic feet in 1825, according to another source; by 1886, "the average selling price of gas . . . was assessed to be $1.77 per thousand cubic feet, but this masked a very wide price range around the country of from 75 cents to $20 per thousand cubic feet."[53] Since the town gas was of very low calorific (Btu) value, the price in MMBtu was often astronomical. Newspapers in the early days of coal gas were outraged at this expense, dubbing the rush to gas light "the heavy charge of the light brigade."[54]

The importance of coal gasification cannot be underestimated. "It lit America's streets and cooked much of the nation's food for many decades. Indeed, a small city wasn't complete without a gasworks, the tangled network of pipes near the railroad tracks."[55] As the industry grew, gas manufactured from coal moved indoors. Around the middle of the century, not just streets but even

> public buildings were lighted in this manner. A few wealthy citizens lighted their homes with coal gas but it was not until the period 1865 to 1875 that the use of gas for home lighting began to make any great progress.[56]

Similarly, in the 1860s, gas began to be used as fuel for cooking as well as for lighting, as the expense declined somewhat.

The last quarter of the century, though, as in England, witnessed the arrival of electricity and the light bulb, and town gas began to be replaced by electric lighting. By 1900, the displacement was severe, and coal's use as a source of light was under heavy attack . . . but coal hung on, to play what the Department of Energy calls "a major role in the United States until well into the 1900s."[57]

During the 1920s and 1930s, the production of gas from carbonized coal—as a source of heat—peaked and crashed. Lighting by coal gas had already virtually disappeared, but around 1920, coal suffered another blow: "the relatively low price of oil led to a switch from coal to fuel oil as feedstock for manufactured gas." [58] By 1931, American production of coal gas totaled only 419 billion cubic feet, less than a third of the overall gas market but still more than any other country in the world. "As late as 1932," DOE reports, "the residential gas market in the Eastern United States was supplied mainly with synthetic gas made from coal."[59] And small, "captive facilities providing low-energy gas for on-site use as industrial fuel involved more than 11,000 gasifiers during the 1920s, with more than half in the steel industry alone.[60] When coupled with remaining town gas plants, the output of these producer gas facilities has been estimated to have reached 2.68 quads by 1949. That is almost equal to 1,300,000 barrels of oil per day. But in that year the transcontinental natural gas pipelines arrived, and coal gas essentially disappeared.

Yet, if the importance of gasification needs to be recalled, so too does the experience of all those years. Operating experience outside England, though nowhere nearly so well-documented, provides some cause for concern. Cancer, explosions, and serious environmental pollution, *have* occurred, with equally serious results. In the summer of 1981, the *Washington Post* reported that small black blobs of coal tar had been discovered floating in a trout stream called Brodhead Creek below Shroudsburg, Pennsylvania. State environmental officials traced the leftover tar to the Shroudsburg Gas Company's gas works, a town gas facility which operated from the 1880s to about 1945. The *Post* reported that:

> This plant used an airless heating process called pyrolysis (or, carbonization), which can leave up to 20 gallons of tarry wastes for each ton of coal it processes. . . . The tar contains several human cancer-causing agents like benzene and naphthalene, plus chemicals like tolulene and xylene that are toxic to fish. The EPA estimates that 8 million gallons of the stuff were dumped into unlined pits or lagoons on the banks of the Brodhead, right at the fork of Shroudsberg's main shopping area. A gravel vein runs through here, and the tar has seeped out to contaminate an underground area of six to eight acres.[61]

The Brodhead apparently has eroded its way down to within six inches of the pollutant lagoons by now, and golf ball size blobs of tar have made it to the surface via natural springs which feed the trout stream. And, noted the *Post,* "it could be a $150 million disposal problem."

No studies exist on American workers, although data from coke ovens (which, as EPA head Costle told the SFC, "are similar to low-Btu coal gasifier[s]") has led NIOSH to term them a "significant health hazard."

The Japanese experience with occupational cancer in "producer" (on-site industrial) gasworks has been conclusive, thorough and horrifying. As Dr. Masatake Kawai recalled in 1967, as early as 1939 an astonishingly high incidence of lung cancer had been seen:

> from 1933 to 1937 among male workers who had had long exposures to tar fume while engaged in the gas generator operation at a steel plant of the Yawata Iron and Steel Works .... [In that period] generator gas was used as fuel in all the steel-making processes at the works. Although a great deal of the gas was produced, and many workers were involved, no precautions were taken to avoid the discharge of tarry fumes into the working environment. The workers were exposed to high concentrations of tarry fumes during their hours of employment.[62]

The death rate went down when the "work load at all plants was reduced . . . [and] strong precautions to avoid exposure of workers" were enforced, between 1946 and 1953; still, 12 more deaths occurred. The initial study suggested a 26-fold excess of lung cancer.[63] The 1967 Kawai study found a death rate 33 times the expected rate, and "workers having the experience of gas generator work for more than 20 years yielded . . . 136 times larger a figure than the expected number of deaths" from lung cancer.[64]

Richard Doll has reported further "positive evidence" of occupationally-caused lung cancer among coal-gas workers in Canada (over 1000% the expected rate) and Norway, where bladder cancer was also traced to work in retort houses. In the Canadian experience, no excess lung cancer was historically found in workers other than retort workers, though, "nor was there any excess among employees at another works where more modern retorts were used."[65]

Overall, then, the Department of Energy in a somewhat understated assessment of nineteenth and early twentieth century coal gas production said:

> the process was expensive and troublesome to maintain, the efficiencies were low, and the operation was inevitably dirty.[66]

Today: How Different?

Today a debate rages about the importance of this historical evidence. In those days, some say we were desperate for gas, and did not know how else to get it. Whale oil and wood were the options then, not petroleum and natural gas, or electricity. Coal we had, and with coal we worked. We were not so aware of the need for environmental and occupational protection as to know how to control the primitive technologies we used.

But today, things are different. So say many in the industry, at least, including Exxon division president Edward David:

> I am convinced that this time, despite the social, economic, and technical challenges, it will be different.[67]

Certainly, some things *are* different today. While technology sat idle in the United States, buried under the flood of cheap, high-energy natural gas, "coal gasification technologies were [being] further advanced abroad."[68] Techniques leap-frogged from simply heating coal in the absence of air, to mixing it with blasts of air, inducing complex reactions in technologically-advanced pressure vessels. This move from *carbonizing retort* to gas generator to *gasifier* kept the European industry alive. As a result, low-Btu gas today provides on-site industrial fuel or chemical feedstocks in scattered places around the world, including the gas for Red Chinese steel mills and a brick factory in Pennsylvania. Elsewhere, on a slightly larger scale, commercial plants gasify coal to produce not only higher grade (medium-Btu, transportable) fuel gas but also "feedstocks" (raw materials) for ammonia, methanol, and petrochemical plants.

The First Generation

New gas technologies for the production of medium-Btu gas appeared between the two world wars to replace the gasworks. To this day only four, so-called "first generation" gasifiers—most importantly Lurgi, along with Winkler, Koppers-Totzek, and Wellman-Galusha—have been successfully commercialized beyond one or two plants.

The important technical transition came with the development of the so-called *water gas* process, by a Pennsylvania professor named Thaddeus S. C. Lowe, in the early 1870s. As Malcolm Peebles suggests in his helpful *Evolution of the Gas Industry,* Lowe's advance "was destined ultimately to outstrip the production of gas by the simple distillation of coal in America," although as we have noted earlier, not necessarily in other countries."[69] The water gas process allowed the gas industry to manufacture more gas, of higher quality, with less waste, and it prepared the transition from retort to gasifier. As Harry Perry has sketched it, the move went like this:

> Gasification involves not only heating the coal, as in distillation, but also the subsequent reaction of the solid residue with air, oxygen, steam, or various mixtures of them . . . Oxygen is expensive, and so the old gas companies resorted to an alternative process to make a gas that supplemented coal gas and coke oven gas.[70]

This process was Lowe's. First, coal or coke was heated in a generator, as in retorting, but then in addition it was "blown" with air to increase the heat and then subjected to a series of "runs" with pressurized steam. The resulting gas burned with a blue flame and a heat value of about 300 Btu/scf. After further upgrading, the gas burned clean and hot. Lowe got a patent in 1873 and the first water gas plant was installed in Phoenixville, Pennsylvania in 1874."[71] (Further refinements in this technique led to a plant in St. Paul, Minnesota, around 1930, where a process was developed which could yield medium Btu gas, which could then be upgraded virtually to natural gas quality.[72] Also called *carburetted water gas,* Lowe's product was important both because its calorific and illumination values could be regulated, and because it pushed gasification beyond simple, expensive, dangerous, distillation. Water-gas plants were introduced around the country in the 1880s and significantly upgraded and cheapened the gas.[73] Lowe's process reached its height at a plant in Billingham, England, where until 1936 some 21 generators turned out Lowe "water gas," used in chemical manufacture.[74]

In Europe, technologies akin to Lowe's set up the move to today's gasifiers. A wide variety of process for making useful substances—not only gas, but chemicals, dyes, fertilizer, explosives, even liquid and solid fuels—from coal became available in Europe, primarily Germany, in the first quarter of this century. Today this variety is difficult to sort out, but the most important were certain coal gas generators (Mond gas producers, Pintsch-Hillebrand gasifiers, and various water gas generators including the "I.G. Farben blue-gas process") and low-temperature coal carbonization ovens (Lurgi-Spulgas and Krupp-Lurgi furnaces). These technologies, long in development (the Mond producer was designed in 1882) provided coal conversion, for the German chemical industry and general industry fuel needs (which we will consider in greater detail in the section on coal liquefaction history).

Then came the gasifiers, vessels in which coal could be heated and reacted with other substances at virtually the same time. Fritz Winkler received a patent for a process which gasified crushed coal in a fluidized bed at near atmospheric pressure in 1922; Bamag Chemitechni began pilot plant tests in 1925 and the BASF complex in Ludwigschafer (near Heidelberg in Western Germany), and by 1928 a commercial plant was running at Levna (east of Leipzig in what is today East Germany). Between 1937 and 1943, three more plants were built in Germany, one in occupied Czechoslovakia, four in Japan, and one in Japanese-held Manchuria.

The German firm of Lurgi finally developed its first real gasifier, a high-pressure fixed-bed unit, in 1936 at a pilot plant in Hirschfeld, and two more plants followed quickly as add-ons to already-large coal conversion (Winkler gasification and Bergius hydrogenation) complexes between 1940 and 1944, at Böhlen near Leipzig and at Brüx (Most) in German Czechoslovakia.

After the war, the technology spread, not like wildfire, but steadily. It never really made it to the United States, thanks to the arrival of natural gas pipelines in 1949, but in countries and places where pipelines were not widespread and coal was readily available, coal gasification established itself. Lurgi provided the gasifiers for the pioneer South African Sasol plant at Sasolburg in 1954, with additions in 1958, 1966, and 1973 (a total of 16 gasifiers), and 36 modern Mark IV Lurgi units anchor the big new Sasol 2 plant at Secunda, with another 36 set for Sasol 3. In Europe, the wartime Czech plant at Brüx (Most) got three new gasifiers in 1949, and between 1955 and 1970 Lurgi built two more plants in West Germany, two in Great Britain, plus one each in Australia, Korea, India and Pakistan. Lurgi technology is also marketed by East Germany (GDR): in 1964 a plant with 24 so-called "PKM-GSP" gasifiers was built as part of a liqnite complex at Schwarze Pumpe, GDR. The East Germans also used the modified Lurgi technology to build a six-gasifier plant at Obilic, near Pristina, in the Kosovo region of Yugoslavia (more on this plant in Chapter 4). (There are also reports of two additional plants in Czechoslovakia, but we have been unable to confirm them.) The plants at Sasol, Schwarze Pumpe, and Kosovo are all running today; some of the others have been shut down and some we don't know much about. American Natural Resources plans to use Lurgi in its Great Plains, ND, plant, the first commerical high-Btu gasification plant in the world.

All but one of the eleven wartime Winkler plants continued to generate after the war, according to Winkler-builders Davy International; two were still going in 1980. Since 1950 six Winkler plants have been built, with two in Turkey and Yugoslavia still running. They also seem to have appeared, says Davy Intl., in the Soviet Union (2) and Bulgaria (2), plus one at the Schwarze Pumpe complex in the GDR (we couldn't verify this).

The third major gasification technology was developed immediately after the war, the entrained-bed Koppers-Totzek (K-T) gasifier. A Koppers unit (which never really worked) was built for the U.S. Bureau of Mines in 1948, and another was ordered for the Mazingarbe Gas Works in France in 1949. The first full-scale K-T plant began operation in Finland in 1951, and since then (counting the French and Finnish plants) a total of 15 K-T plants have been built, in Japan, Spain, Belgium, Portugal, Greece, Egypt, Thailand, Turkey, East Germany, Zambia, India, and South Africa. According to manufacturer Krupp-Koppers, these fifteen plants have involved 54 gasifiers, 47 of which use coal as raw material (oil or oil products like naptha can also be gasified in a K-T unit). As of January 1979, seven of these plants, *all coal-using,* were still running, all of them used to turn coal into synthetic ammonia.[75]

The recent history of coal gasification has involved much research and some development at the pilot and demonstration plant scale, but little commercial activity. The generator era came and went within about 20 years, thanks to natural gas which flooded the world in the 1950s. The Mond producers and the Lowe process, along with the handful of other water-gas processes, disappeared about the time of the Korean War. The Winkler units seem to have hung on and the Lurgi and Koppers-Totzek gasifiers made it on to the scene just in time to survive, but too late to make a big difference: just when "the technology had reached a stage where plants could be installed, natural gas was discovered in the North Sea and in North Africa . . . . Few coal gasification plants embodying new technology were installed, and interest in further improving the technology flagged."[76] Today, the old town gas works have finally closed down, and no commercialization of second generation gasification technologies has occurred. This history is best summarized in a report submitted to NIOSH in April 1979: "somewhat mixed," varying "indirectly with the availability of natural gas and petroleum."[77]

No commercial plants producing high-Btu gas (pipeline-quality, synthetic natural gas, or SNG) exist; so far, SNG technology has advanced only as far as the demonstration stage, at a British Gas plant in Westfield, Scotland, and a handful of pilot plants in the United States.

What about the "difference," then? There are no longer 11,000 gas units, as there were in the 1920s. Today, coal gasification could not begin to come close to its 2.68 quads of 1949. But the commercial plants which do exist should be safer, cleaner, and comparatively cheaper than their gas works counterparts. On the workplace safety and health front, the argument has some validity. One scientist with a British gas board, involved in the Doll gasworker studies, is reported in NIOSH documents to make this sort of claim, arguing that on the basis of his experience he[78]

> does not see any particular significant medical problem with the [Lurgi gasifying] operation as opposed to the coal carbonization process on which he has reported. Those gas workers used a retort carbonization process, and were found to have an increased prevalence of lung and bladder cancer.

And NIOSH has agreed as well:[79]

> Gasworks, where coal is heated in retorts for the Primary purpose of producing flamable gas, . . . heat coal with minimal containment of the volatiles. Gasification plants typically operate at high pressure as well as high temperature, and therefore containment is essential under normal operating conditions.

Thus, NIOSH concludes, although "epidemiologic studies of gasworkers . . . *have* provided evidence of work-associated cancers,. . . this evidence does not imply comparable degrees of hazard in coal gasification plants [and] gasworks." Yet, the report continues, this does not mean that the *potential* for danger is at all lessened:[80]

> The nature of the toxicants and their concentrations within the various process streams at future coal gasification plants are expected to be *generally similar* to those at gasworks . . ." [emphasis added]

NIOSH places its faith in "proper design and work practices which it promises" can substantially reduce the quantitative aspects of exposure. The question becomes one of containment—of housekeeping—the old plants were dirty, because no one cared or knew enough to keep them clean, and that failure accounts for the death, disease, and pollution noted there. But in gasifier operations, as opposed to retort works, the plants are less dangerous, because of the attention given to cleanliness and safety.

This discussion has touched only on in-plant worker exposure to cancer-causing volatiles, but there are other hazards and considerations. The effects on people in neighboring towns or cities, or impacts on the local environment, have not been considered; nor has the historical experience of unreliability or the high cost of both plant and product or other non-carcinogenic hazards of the workplace. Presumably, today's plants and those of the future will be different, but there *is* reason to be skeptical. Modern gasifiers *do* in fact have an operating history of their own, above and beyond the experience of the manufactured- or town-gas works, and, although it is sketchy, it offers plenty of suggestions, and plenty of warnings. The question is whether the hazards will be acknowledged and controlled.

LIQUIDS FROM COAL

Just as the medieval alchemist sought to turn base metals into gold through the use of the philosopher's stone, so have modern scientists sought for the last seventy-five years to turn coal into oil. Once naturally occurring liquid petroleum became widely available for industrial uses, its superiority over coal was obvious, even to the most ardent coal partisan. Oil could be piped, stored, refined, altered and controlled to make it fit a given industrial need in ways that coal could not begin to match. But oil is a fickle commodity, spread unevenly about the world. One never knows where it will be found or when it will run out. Coal is widespread. And since the turn of the century, coal-rich, oil-poor nations have looked for ways to convert their great black coal deposits into liquid gold—oil. As this section will show, there have been technical successes at doing this, but they have always ended in economic failure. The process has always been fraught with difficulty and danger.

"Coal Oil"

Attempts at coal-to-oil alchemy may date back as far as 1790 when, by one account, synthetic oil was used for "cooking, lighting, and heating."[1]

More reliably, we can date the emergence of coal-to-oil technologies at around the middle of the nineteenth century in North America and Scotland. Early processes used *cannel coal* or *bog-head coal* which has recently been pinned down as "a peculiar substance with a composition somewhere between a very high-volatile bituminous coal and a high-grade oil shale."[2] Scottish scientist James Young commercialized a process for lengthy, low-temperature distillation which recovered paraffins and oil from bog-head coal in 1850. The process was much like that used to pyrolize coal for the production of coal gas. By 1857, Scotland is said to have had a "substantial coal oil industry" involving 50 plants producing 20,000 gallons a year of kerosene-like lamp oil from the distillation of coal, shale and natural asphalt.[3] Young's process was attractive because it produced a cheap replacement for whale oil which by mid-century cost up to $2.50 a gallon.[4] In the same year that Young built his first coal oil plant, Canadian Abraham Gesner discovered that heating New Brunswick coals (really more like asphalt or tar sands) yielded very high-grade oil and gas. For reasons unknown to us, "Canadian courts prevented him from developing [this] coal."[5] But the process caught on in the United States, during the decade of the 1850s, flourishing for about 10 years—like the domestic oil shale industry—until drowned in the flood of Pennsylvania crude oil.[6]

While the industry lasted though, plants thrived along the Eastern seaboard, notably in Pennsylvania, West Virginia, and North Carolina. Plants on the Ohio River in Kentucky are said to have reached a production rate of 13,000 gallons a week in the last years of the decade. Hammond and Baron report that:[8]

> The Downer organization of Boston, the leader of the industry was producing coal oil at the rate of 900,000 gallons per year by 1858. Recovery rates were as high as 110 gallons per ton of coal ... [This] coal oil cost approximately $7/million Btu [$42/barrel] in 1860 dollars. Given the uncertain status of the raw material used in making coal oil—"cannel coal" can refer to both a coal-like shale and a shale-like coal—the history recounted here overlaps considerably with that of the early American and probably Scottish oil shale industries.

"Light Oil"

The process involved in coal-oil manufacture was what we now call "pyrolysis" or retorting, coking, carbonization and (destructive) distillation.

It is the basic hydrocarbon conversion method, as the histories of coal gasification and oil shale demonstrate. When you heat coal in the absence of air, you get *combustible* products, not the combustion products of burning (heating in air). As we have seen earlier in other synfuel processes, pyrolysis developed as a means of making coke for iron smelting, in the eighteenth century, and gradually uses were discovered for the other products of coal carbonization—gas and tar. For a while, then even oil was directly recovered, until its role in the growing industrialization of the West was taken over by crude oil. But just as coal gas hung on in spite of petroleum, so, too, did coal liquids. When synthetic oil was deposed by crude, the owners of the carbonizing ovens shifted to the production of other important hydrocarbons for *non-fuel* uses.

This transformation focused on the recovery of condensable materials after carbonization. Originally, these "were either burned or sold for use as patching materials, etc. As the art of distillation improved, the condensable liquids were separated into fractions such as benzene, anpthalene, creosolic acids and the heavy tarry residue."[9] That "heavy tarry residue" became the basis of the synthetic dyestuff industry. As for the rest of the condensables, they were known as *light oil.*

Light oil recovery got its start at the Bradford Road Works of the Manchester (England) Gas Company in 1881. Product gases were "washed" with a creosote oil, causing the condensation of benzene, an important chemical. Benzene, along with its chemical relatives tolulene and xylene (the so-called "BTX" family), has a tremendous range of industrial and chemical uses and was widely sought after. In fact, synthetic BTX-chemicals are said to have "formed the sole basis for the organic chemical industry prior to World war II."[10]

From its start in Manchester, this technique of "light oil recovery" developed into the basis of an industry of considerable strategic importance and profitability. A plant devoted to making BTX chemicals from coal was built in Syracuse, New York in 1898—as an add-on to a battery of coke ovens—and by 1914:[11]

> there were 14 light-oil plants, all of them operated by one company. WWI brought with it a high demand for BTX chemical products, and in 1915 the number of light oil plants had doubled to 30. Coke ovens and gas-works supplied more than 80% of the light oil for organic chemicals during the war; by 1922, though, it was recovered at only two gas works.

The end of the war brought with it another petroleum rush and coal was once again replaced as the feedstock. As the light-oil business was developing in the United States, another similar hydrocarbon conversion industry was growing in Europe, especially in Germany. Capitalizing on a revolutionary discovery—a discovery that had proved resistant to commercial exploitation in England—German industrialists set about developing a synthetic dye-from-coal tar industry. Germans in England exported the previously unsuccessful process back to their homeland and, in the words of historian Joseph Borkin, whose *The Crime and Punishment of I.G. Farben* provided most of the following narrative:

> what they did with their booty was nothing less than an industrial miracle. With the German knack for turning garbage into wealth, these talented borrowers transformed the mountains of coal tar, the costly waste of the steel production of the Ruhr, into an immensely valuable product, the raw material for a new and exciting dyestuff industry.[12]

This meant that coloring, for the first time, no longer depended on natural colors. Of course, synthetic dyestuffs were immensely profitable, and most of that wealth came to be concentrated in a small group of German chemical firms which immediately dominated the world dyestuff market and turned to investigating other coal products. This spectacular success in making dyestuff from coal tar led to bigger projects culminating in the Nobel Prize-winning discovery of nitrogen fixation in 1909 and, within four years, the actual production of synthetic ammonia.

The ammonia plant, which was built at Oppau by a firm called Badiche Anilin und Soda-Fabrik (BASF—the largest of the dyestuff companies) was something of an industrial miracle. The conversion process utilized extremely high temperatures and pressure, as well as stable catalysts. The plant was built with superstrong alloys so that the equipment could operate under the severe conditions. BASF reaped "immense financial returns" and the stage was set for the first sophisticated venture into the production of coal liquids.[13]

Coal Liquefaction

Having figured out both the chemistry and the engineering of converting coal to ammonia, the dyestuff industry seemed secure and profitable. But just when the financial future seemed brightest, World War I intervened and military allocation decisions brought the ammonia-dyestuff industry to a standstill. In a move which has come to characterize the coal-derivative industry, BASF turned to the government, offering first a new synthetic nitrate process for gunpowder, and then an idea for producing—this time on purpose—poison gas from coal. [14] By the summer of 1915, BASF had convinced the government to build a huge syn-nitrate complex at Leuna, where it "produce[d] almost nothing but military contracts."[15] Soon the bonanza began to spread to the rest of the dyestuff industry and—even though Germany lost the war—the industry found its footing again. The crucial lesson had been learned: stick with the government. Yet another rule had also become apparent: stick together. The vision of a possible German defeat had brought most of the chemical industry together in a loose cartel, an *Interessen Gemeinschaft,* or "community of interest," in 1916.[16] And along with the I.G., the war brought into being the first serious attempts to liquefy coal for fuel. For the next 30 years, the four would always be linked: the I.G., coal liquids, the German government and war.

The first work on advanced techniques for turning coal into oil began around 1910 in two different laboratories, both linked to the I.G. In 1911, scientists M. Pott and W. Broche started to develop a process known as *solvent refining,* in which an oily solvent is brought together under pressure with coal to dissolve it into liquids. Development continued throughout the war and during the twenties eventually shifted to BASF/I.G. where it was demonstrated commercially at the Leuna plant. The first uses of solvent refining, though, were not as fuel: "the dissolved substances were not separated from the solvents, which were used for a long time as a cheap newspaper printing ink."[17]

The development of the Pott-Broche solvent refining did not seem to have much commercial viability and it was easily overshadowed by a later I.G. discovery. During the war, a chemist named Friedrich Bergius—working under contract to BASF in the lab which had developed the synthetic ammonia process—continued experiments which he had begun around 1907 into the use of "catalytic hydrogenation" to produce coal liquids. In 1914, Bergius took out a patent for a process, by means of which:[18]

> coal, treated with hydrogen [and a throw-away catalyst] at high temperatures and high pressures yielded 85 per cent of its moisture-ash-free weight in soluble or liquid petroleum-like hydrocarbons.

In 1916, Bergius began attempts "to adapt his hydrogenation process to large-scale production. However, he had still not succeeded by the end of the war."[19] That success was to belong to the I.G.

The crucial year was 1925. BASF, as Borkin tells the story, felt the impact of two important events: the "clandestine rearmament" of Germany had begun and BASF's "domination of the world's synthetic nitrate industry was coming to an end."[20]

Commercialization of the Bergius process could, at the same time, assure a new German army of secure and unnoticed fuel supplies and eliminate the need "to shut down a large part of the costly, high-pressure installations at Leuna and Oppau."[21] But the acquisition of the rights to the Bergius process, not to mention the staggering financial requirements for commercializing it were well beyond BASF's resources. The time was ripe for a merger. In December of 1925, the seven other members of the I.G. cartel merged themselves into BASF and formed *I.G. Farbenindustrie Aktiengesellschaft,* immediately "the largest corporation in Europe and the largest chemical company in the world.[23]

Even before the I.G. Farben was formed—but when the merger had been all but consummated—BASF had purchased the Bergius patents. Hydrogenation and Farben seemed a perfect match. Not only would a synfuels effort provide a new and profitable use for the expensive equipment at Leuna and Oppau, but the merger "arrived in the midst of the great European auto boom and apocalyptic announcements of the imminent exhaustion of the world's oil reserves."

New Farben leader Carl Bosch set in motion the Farben research machine: "Before long, work was underway to adapt the Oppau plant from nitrate synthesis to the conversion of coal to oil."[25] He also set out to raise capital for the plan. The first target was Standard Oil of New Jersey (Esso, now Exxon) whose executives were so frightened by the prospect of a European coal liquids industry—said one, after seeing Oppau, "this means absolutely the independence of Europe in the matter of gasoline supply"[26]—that Bosch knew he had a gold mine.

In June 1926, Bosch announced that the Oppau experiments would be scaled up immediately in a new 100,000 ton/year Bergius plant to be built next to the Leuna ammonia plant. Within a year, though, the initial euphoria dissipated, as the plant became bogged down in a morass of troubles, not unlike those which have come to be associated with modern synfuel efforts:[27]

> The Bergius plant at Leuna, which had begun production in June [1927], was beset by operational failures and extremely serious technological problems. Expenditures had soared so far beyond the original that, if continued, they would threaten the financial structure of the I.G. Farben itself.

The immediate solution to the cash crunch was Esso, and by 1928 both companies got what they wanted. Esso got the Bergius patent rights for the entire world (except, of course, Germany) and thus control of European energy independence: one Standard official put it, "The I.G. are going to stay out of the oil business—and we are going to stay out of the chemical business." And Bosch's I.G. got $35 million, enough to keep Leuna afloat.

If Esso ever really had any plans to commercialize Bergius hydrogenation, they disappeared quickly when huge new oil finds in Texas ended the petroleum crunch which had made coal liquids attractive. At any rate the twin problems of shortage and competition disappeared. Farben's problems, on the other hand, refused to go away. The Leuna plant was "drowning in its own economics . . . the cost of [its] synthetic gas . . . was forty to fifty pfennigs a litre, whereas the world price of natural gasoline was only about seven pfennigs."[28] Farben soon needed another bail-out. This time the source was Adolph Hitler. At about the same time that Bergius and Rosch were being rewarded for their hydrogenation efforts with the 1931 Nobel Prize in Chemistry, initial contacts were made with Hitler, who headed the up and coming Nazi party. At a meeting with Farben executives in November 1932, three months before he became chancellor, Hitler insisted that "German motor fuel must become a reality, even if this requires sacrifices. Therefore, it is urgently necessary that the hydrogenation of coal be continued."[29] In March 1933, Hitler and Bosch met, and the new leader "gave Bosch absolute assurance that his government would fully back the synthetic oil project, and Bosch agreed to expand the Leuna plant."[30] Farben's troubles were over and in December the first price guarantees for coal liquids were agreed upon:[31]

> By terms of the contract I.G. was to expand the synthetic oil installation at Leuna so that in four years, by the end of 1937, it could produce

> 300,000 to 350,000 [metric] tons annually. The Reich, in return, pledged to guarantee a price corresponding to the cost of production, including a five percent interest on invested captial and generous depreciation, and to take measures to assure the sale of all synthetic oil not sold by I.G. through its own outlets.

In the Nazi four year plan for war preparation, drafted by Hitler in 1936, I.G.'s coal liquids were central; "just as we produce 700,000 to 800,000 tons of gasoline at the present time, we could be producing 3 million tons."[32] This was the final boost for Farben and coal liquids and the hydrogenation business expanded accordingly. By the end of 1936, Farben plants had turned out over 1.4 million metric tons of gasoline, and the figures continued to rise steadily until the end of the war.[33] By early 1944, in fact, twelve Bergius plants were producing at a rate of more than 3 million metric tons a year. Thanks to the Four Year Plan, a third synfuel technology came into the picture in 1936. This was a much simpler, and thus easier to commercialize, process, based on the old water gas plants. It had originally emerged in 1925, when[34]

> Franz Fischer and Hans Tropsch of the Kaiser Wilhelm Institut fur Kohlenforschung in Mulheim, Ruhr, published an account of their synthesis of a mixture of liquid hydrocarbons from water gas at temperatures in the region of 2000 degrees C, at atmospheric pressure, in the presence of catalysts.

The Fischer-Tropsch process differed from solvent refining and Bergius in that it involved two steps—first the production of coal gas, then the molecular rearrangement of that product into liquid fuel. This discovery was important, because it provided the solution for a pressing problem in the Ruhr—overproduction of coke by Germany's steel industry:[35]

> A mountain of coke had formed in the Ruhr and, as a result of research with the object of finding a use for it, it was found that liquid fuels could be produced by means of gasification of the coke. So the Germans initially gasified this material, thus making use of coal only in an indirect [Fischer-Tropsch synthesis] way.

Once again, German technology made the leftovers useful, and, by 1935, construction of the first Fischer-Tropsch indirect liquefaction plant began, after extensive research on commercialization had been done by Fischer and Tropsch's industrial benefactor, Ruhrchemie A.G., the chemicals branch of the Ruhr coal syndicate. In 1936, the first Ruhrchemie F-T plant began operation and with the help of the Four Year Plan—with its generous financial incentives for the chemical industry and its stiff import tariffs—eight more Fischer-Tropsch plants were producing liquid fuel by 1939.[36]

World War II

On the eve of World War II, Germany had a working industry producing a little more than 2 million metric tons (about 15 million barrels) of synthetic oil per year. As we have seen, four technologies were its base, and to this day innovation has not pushed us any considerable distance beyond them:

1. *Simple pyrolysis or carbonization, destructive distillation or coking.* The heating of coal in the absence of air to yield by decompositon hydrogen-rich gases and liquids along with a solid carbon residue or coke.
2. *Solvent refining.* The extraction of a hydrogen-rich liguid solution from coal after it has been mixed under high temperatures with a specially selected solvent.
3. *Catalytic hydrogenation.* The production of coal liquids by the addition of hydrogen to coal in the presence of a catalyst under high pressure and temperature.
4. *Indirect liquefaction.* Production of coal liquids by catalytic rearrangement of the products of coal gasification.

The brief survey below shows that each of these technologies was in use in pre-war Germany. The survey also indicates the level of development in other countries.

Carbonizing ovens were already widespread, using both direct and indirect heating furnaces. In 1937, over 600,000 tons of low temperature lignite tar were produced for the dyestuff industry. Tar yields were high. Indirect furnaces could handle between 10 and 35 tons of coal a day; direct furnaces could handle upt to 250 tons per day. The average carbonization complex might contain as many as fifty furnaces.[37] As the coal liquids effort intensified, these carbonizing ovens began to provide tar as a feedstock for hydrogenation plants, as well as producing benzole and large quantities of fuel oil directly.

The Pott-Broche solvent refining process never made it much beyond the boundaries of the Leuna plant into which it had been integrated in 1927. One plant was built in Japan during the war to produce oil, but the main use of solvent refining was to produce a feedstock for the Leuna hydrogenation plants.[38]

As noted above, Farben constructed 12 Bergius plants. This process also was used outside of Germany. In 1931, Farben had expanded its relationship with Esso to form an international cartel to control all of the patents. Called International Hydrogenation Patents, Inc., it consisted of Farben, Esso, Royal Dutch Shell and International Chemical Industries (ICI) Ltd. Over 3,000 patents were shared. Esso applied the technology to crude oil and discovered that it could double gasoline production without increasing the crude feedstock—this was one of the first major steps in modern refinery improvements. Although Standard built no coal plants, the other partners did. ICI built a plant in Billingham, England in 1935 which ran successfully on coal and coal tar from the ICI ovens for five years, until creosote replaced coal as the feedstock. Production at Billingham was 15,000 tons per year. Two more plants utilizing coal tar hydrogenation were built in Italy in 1936; they had an output of 240,000 tons per year. Another plant was built in Korea in 1942; it produced 110,000 tons of gasoline and diesel fuel a year.[39]

As for the coke eating Fischer-Tropsch process, a total of nine plants were built in Germany by 1939, all under license from Ruhrchemie with an aggregate rated output of 740,000 tonnes per year—about 16,000 barrels per day or 5½ million barrels per year. These plants were quite primitive. Synthesis gas came not from the first generation of gasifiers then being developed but from older "water gas" generators and gas purification—what there was of it—was a "complex affair, comprising a number of process stages, and very expensive."[40]

But regardless of their state of technological advancement, the plants spread. Besides the nine built from 1936-9 (none thereafter), plants were built outside of Germany: one in France, one in Manchuria and two in Japan.[41] Ruhrchemie also sent Dr. Fischer to South Africa in 1938 to spur interest in a plant there, but no arrangement could be completed.[42]

Which brings us to World War II. It should be apparent from the preceding lengthy history that, in fact, "there never was a crash wartime program to build synthetic fuel plants and Germany built less than one quarter of its synfuel plants during the war," as one observer has put it.[43] At its height, in 1944, the German synfuel industry consisted of the nine Ruhrchemie F-T plants and the twelve Farben Bergius plants, along with scattered carbonizing ovens in dye factories and the SRC unit at Leuna. Most of these plants were built during the decade-long period before the war broke out. The total rated production of all German plants was 16,000 metric tons per day—about 120,000 BPD—and maximum actual production rate peaked at around 100,000 BPD. The largest of the plants was rated at 12,000 BPD.

And even while synthetic fuel processes were producing at their maximum rate of 4.8 million metric tons per year into the war, total German petroleum production reached a yearly rate of almost 15 million metric tons. Germany did not run the war on synthetic fuels: there was a thriving conventional petroleum industry at work at all times. The tremendous Allied air strikes of April-May 1944, which reduced the synthetic fuel industry to rubble, worked to end the war not because they destroyed the coal liquids plants—although that certainly helped—but because they levelled the oil refineries.[46] The plan to convert the economy to coal liquids—to use, in Hitler's words, "solution of our necessities"[47] was a failure,[48]

> despite the legacy of World War I, a head start in developing the technology, the organized might of the Nazi state behind an ambitious conversion plan and the chilling imperatives of real, rather than economic, warfare.

And despite I.G. Farbenindustrie, the largest corporation in Europe, stocked by the world's best scientists and engineers, with a 30-year program of research and development on a scale the likes of which the world had never seen undertaken before.

The plants were built only with massive subsidies. None of the plants was meant to run at a profit and if the initial experience with Leuna is any indication, their unprofitability was enormous: with the 1933 price guarantees the Third Reich agreed to pay *seven or eight times* the price of petroleum-based gasoline for coal liquids.[49] There were, of course, big profits for I.G. Farben, but they came from the government, not the process. The process, in fact, was not only "extremely expensive," but also horribly inefficient and difficult to manage; as the U.S. Department of Energy has noted, the German plants were "complex, bulky and expensive."[50]

Complex, indeed. In spite of Bergius's and Bosch's combined technical genius, the Leuna plant was plagued with operational failures, technical mishaps and huge cost overruns. "Environmental protection was not a serious concern" in the wartime plants, and according to *New Scientist*, data from the plants "indicate that the cancer incidence was many times higher for coal conversion workers than for others."[51] Technical documents on the operation of the plants show that the chief concerns in coal liquids plants were production and protection from air attack: occupational safety handbooks detail extensive measures for shelter during air raids, but no new measures beyond the minimal personal hygiene required of workers in pre-war dyestuff plants.[52]

After the War: United States

After the war, American, British and Soviet military and scientific personnel went through coal liquids files held by the government, Farben and Ruhrchemie. The American team, called the Technical Oil Mission, came home to the Bureau of Mines Synthetic Liquid Fuels Division and set to work using captured documents and new knowledge to implement the Synthetic Liquid Fuels Act of 1944 in order to encourage and develop a domestic coal-to-oil industry, train engineers and scientists and determine whether American equipment and coal could replicate the German experience. In 1947, the project obtained the Army's Mission Ordnance Works—a wartime ammonia from natural gas plant—located in Louisiana, Missouri.[53] In May of that year, they began to convert the plant into a Bergius facility. In March of 1948, construction began on a Fischer-Tropsch plant at the same site.

Operation of the Bergius plant began in April 1949, after a lavish dedication ceremony on the anniversary of V-E Day, at which "diesel fuel produced during the first run was used to fuel a diesel electric locomotive that hauled a loaded, 8-car special train from Saint Louis to the plant and back. The round trip was approximately 200 miles."[54]

The plant, rated at 200 barrels a day, ran intermittently until 1953, testing American coal and American know-how. In 1952, some 10,000 tons of coal and lignite went through various stages of the Bergius plant and the year's "finished commodity" output amounted to "435,000 gallons of 76- to 77- octane gasoline, the bulk of which was shipped for use in military fleet tests, and a smaller amount was used in demonstration plant vehicles."[55]

Overall operations in the hydrogenation plant—which lasted from April 1949 to June 1953—were said to be quite successful, at least in terms of the project goals. The process seemed to work, sometimes it actually produced usable liquid fuels, and when it did not, problem areas were identified. And of course, problems did arise, some of them quite serious. Looking only at the operations for 1952, its last complete year of operation, the list is lengthy: instruments and controls periodically became inoperative on cold days; heavy deposits of coke and solids built up in heat exchangers, preheaters and converters, necessitating expensive cleaning; centrifuges and pumps failed, often requiring that some process stages be simply by-passed; flanges, gaskets and valves failed, producing dangerous leaks, and so on. To these routine problems which seriously impacted efficiency were added two more serious incidents. During one run, a short circuit in the process caused temperatures in the converters to rise quickly to over 1200°F, 50% higher than normal and the risk of an explosion was averted only by dumping the converters and injecting cooling gas. On a second occasion, the plant was not so lucky:[56]

> After approximately 24 hours "on stream" [during one run], it was necessary to shut down the coal preparation plant owing to conveyor trouble. During shutting down an explosion occurred in the duct and cyclone system, which damaged the cyclones, duct work, and the upper part of the building seriously.

All this in one year, just three runs, at a "successful" plant. The second plant at the Louisiana Missouri site was a gas-synthesis or Fischer-Tropsch unit. Construction began on this plant in March 1948, and operations began in September 1951. Like the hydrogenation facility, it was run only a few times over a two year period and was plagued with frequent breakdowns and accidents. Though all sections of the plant operated successfully at one time or another, the bureau could never get them to work together. So while the Koppers gasifier (an advance over those used in the Ruhrchemie plants) did actually produce coal-gas, the Fischer-Tropsch unit never utilized it and instead relied on coke-based producer gas as a feedstock.

There were several problems with the gasifier. It never achieved required carbon conversion rates. And it had a tendency to disintegrate; at one point the entire inner lining was burned away and during another run, the injection of raw materials nearly destroyed the entire unit. Finally, a brand new gasifier had to be brought in to replace the original unit.[57]

As for the Fischer-Tropsch synthesis unit itself, the Bureau of Mines had these cautionary words:[58]

> In operating any new processing unit, it is not unusual for mechanical troubles to develop, and it was not unexpected when some of the equipment in the synthesis and distillation units . . . did not function properly.

In 1952, only one synthesis run was completed and it had to be terminated abruptly when dangerous leaks developed after just two weeks.

In March of 1953, after they had been "carefully reviewed by Bureau [of Mines] and industry technologists," both of the plants were ordered to shut down permanently.[59] By May or June of that year, this had been accomplished and, in September, they were once more in the hands of Army ordnance. The explanation provided by the Bureau was that the project's main objective had been attained, that is, converting American coals to oil with American staff and equipment had been shown to be "technically feasible."[60]

Just as the government program was shutting down, a private effort at coal conversion was beginning in the hollows of West Virginia. In May 1952, Union Carbide began "the first large-scale pilot plant production of chemicals from coal hydrogenation"[61] on a 20- acre site at Institute, West Virginia. The plant, aimed not at producing synthetic fuels but petrochemical feedstocks, ran until 1956 at full-scale (300 tons of coal per day), and until 1962 at a greatly reduced level (12 TPD), as Carbide apparently became progressively more and more disenchanted with the commercial prospects of separating the chemicals from the rest of the output and as workers became more and more sick.[62]

In 1955, after reports that materials in some process streams caused cancer in animals, quarterly skin examinations were begun, and human cancers came to light immediately. A fairly rigorous industrial hygiene program was quickly instituted, but despite this, inspections over the next four years revealed that 50 of the 359 workers examined had developed cancerous or precancerous skin lesions. This rate of occurence was 22 times that of people in the rest of West Virginia and between 17 and 37 times the national rate (adjusted for sex, race, and age). Carbide's medical director "concluded that an increased incidence of skin cancer was found in workers exposed nine months or more to contact with coal hydrogenation chemicals."[63]

A follow-up study, conducted in 1977 and tracing the 50 workers with lesions, found no "increased risk of developing systemic cancer," had resulted from the exposure, but conceded that "the possibility of systemic cancers still exists" thanks to "glaring epidemiological shortcomings" in Carbide's reports.[64] The simplest of these was that no one has ever traced the 309 workers who had not developed lesions prior to 1959. In computing the statistical significance of the higher than average occurrence of cancer among the 50 with lesions, it is assumed that none of the other 309 ever had any problems—a less than likely outcome—and their assumed zero rate lowers the second study numbers below the statistical significance level. And the fact remains that one of every seven workers in the Institute plant developed skin lesions before 1959—that none of them have yet died of it (which is what the second study really examined) provides only scant consolation.

The federal government began subsidizing coal liquids efforts again in 1961 through the Office of Coal Research. OCR sponsored construction of a small pilot plant in Princeton, New Jersey,[65] and construction by Consolidated Coal Company of a solvent refining plant in Cresap, West Virginia, located near a Consol petroleum coke plant and the Kamerer power plant which Consol had in the early 1959 "persuaded Appalachian Power to construct so that it could burn char produced by low-temperature carbonization" in Consol's coke plant.[66] The Cresap coal liquids plant was a disaster and it was shut down and mothballed by OCR in February 1970. "The operation of the pilot plant was fraught with [serious technological] problems, . . . in extraction, in temperature carbonizer, solid-liquid separation and hydrogen section."[67]

OCR continued to find coal research projects to fund until into the 1970s, but no other plants were constructed. Its budget was too small for the size of even pilot plant proposals. One project which got started with OCR money, but took 12 years to come to fruition—and thus outlived OCR—was Solvent Refined Coal, which had its genesis in a 1962 grant to the Spencer Chemical Company for Project Low Ash Coal. Today, two versions of the SRC process, along with H-Coal (also funded at one point by OCR) and the Exxon Donor Solvent process have all been built—with federal funding—at the pilot plant stage. These projects and their histories are discussed at some length in Chapter 2, which examines present technical processes. All of these projects were funded by the Energy Research and Development Administration (which absorbed OCR in a government reorganization) or its successor, the Department of Energy.

Sasol

Like the German wartime plants, South Africa's Sasol plant has acquired considerable notoriety, but it has not produced much fuel. Sasol provides a relatively limited amount of coal-based liquid fuel. The Sasol complex (which includes the liquefaction plant) is devoted primarily to the production of petrochemicals, many of them petroleum-based, and the refining of petroleum itself. Like the German plants, Sasol required a long genesis and heavy government aid; its history has featured explosions, pollution, heavy expenses, a repugnant racial policy . . . and, in the end, successful production of a relatively small amount of liquid fuel.

Sasol's history has been admirably and admiringly documented by Johannes Meintjes in a book called *Sasol 1950-1975*. As Meintjes tells it, Sasol began, as we discovered earlier, with a visit to South Africa in 1938 by Dr. Franz Fischer. He came to encourage the Anglo-Transvaal Consolidated Investment Co. Ltd. (Anglovaal) to build a Fischer-Tropsch plant; Anglovaal had been considering such a move for some years.[68] The war intervened and any plans were shelved. But in November 1945, Anglovaal asked the South African government for price-protection supports.

Anglovaal finally got the license it sought—after several tentative approvals and hesitations—in 1949, but found it difficult to proceed, because it did not have sufficient free capital. Anglovaal owned the very rich coal fields near which the plant was to be sited and had already devoted considerable effort to the project, so it was understandably unwilling to give up entirely. It sought increased government involvement. After some debate, the South African government adopted an energy policy in which a coal-to-oil project became "National Priority No. 1." [69] And in September 1950, a fully government-sponsored corporation, the South African Coal, Oil and Gas Corporation, Ltd., known by the acronym of Sasol, took over the project. Backed by funding from the government's

Industrial Development Corporation, Sasol was underway. Construction began next to the Anglovaal coal mine (called Sigma) in mid-1952, and the plant was completed in August 1955. The plant, now called Sasol I, was joined as it rose by a company town named Sasolburg. The town has since become a source of national pride, a masterpiece of central planning; Sasol I, on the other hand, ran into difficulties early—already, in 1954, Sasol managing director Etienne Rousseau was sounding cautionary notes:[70]

> The giant which we have built and are still building has to be domesticated into a purring animal. To courage and perserverance have to be added deliberation and meticulous care and thoroughness if we are not to have a wild beast on our hands . . .

But as 1954 progressed, Rousseau apparently had little to worry about: coal mine, power plant, oxygen plant, and gasifiers all started up with success. [71] The Fischer-Tropsch units, half of them American, half German, were another story altogether. It required two and a half weeks of continuous work just to get the American designed Synthol units going—valves cracked, heaters failed, "unexpected complications in the ash removal and gas feed systems were experienced," and the entire unit came close to disintegrating—and even when they were finally turned on in August 1955, the problems had only started.[72]

A month later, the German Arge units joined in, but the illusion of success was short-lived. As Richard Myers has written, the synthesis technology "proved somewhat tricky and sensitive to operate," so that Sasol "went through an extremely painful shakedown period lasting for more than five years."[73] The Rectisol plant, the first of its kind in the world, needed time. The Arge reactors would not work until mechanical features had been changed. The power plant and the wax processing plants both needed major modifications as time went by. But most seriously, the Synthol synthesis unit—the heart of the plant—quite simply "refused to work as it should:"[74]

> The process was supposed to be continuous: it should have been possible to change the catalyst without stopping the process. This proved to be impracticable, however, and it was decided to change the operating procedure. The two reactors were then filled with catalyst, put into operation, and in turn taken out of operation when the catalyst had been spent . . . The performance of the imported catalyst also left much to be desired, and a great deal of toil went into solving the problem.

These problems may not sound overwhelming, but they were not solved until 1960, and only then by the installation of an entirely new Synthol unit. Sasol made no money for those five years and came under heavy fire in South Africa as a gigantic boondoggle. As chief Rousseau remembers, the first five years were indeed [75]

> extremely painful . . . had Sasol been standing in any other country, . . . I mean in a heavily industrialized country, people would have certainly walked off their jobs. They would not have seen the thing through. But we had a virtually captive staff and they *had* to see it through . . . I recall [people] coming to see me at half-past five one afternoon, having had three failures in getting a unit to work that day, and . . . saying to me, "Etienne, do you think we're any good?" What a battle it was! . . . I must tell you honestly that there were times in Sasol's early years, times when we had trouble, big trouble, when I felt that my main charge was to keep up the courage of our men.

Or, as another Sasol official put it, "We had more trouble with the plant than with any other undertaking I have ever heard of."[76]

Once they got the plant working, Sasol I did not concentrate on producing fuel. Instead Sasol produces around 5,000 barrels of liquid products, mostly petrochemical feedstock; a small operation in modern terms.[77]

The Sasol emblem of a little man bearing a checkered flag adorns gas pumps throughout South Africa and the stripe says "Power from Coal," but Sasol built a crude oil refinery in 1971 and its fuel output bears the Sasol name.[78] Since 1962, the work at Sasol I has been consciously shifted towards chemicals, and though Sasol I itself remained quite small, a vast industrial complex grew up around it, including three fertilizer plants, a plastics factory, three synthetic rubber plants, the 55,000 BPD Natref oil refinery, a naptha cracking plant, a butadiene factory and a styrene plant.[79]

In 1974, Sasol announced plans for a new plant, Sasol II, to be located next to a new town to be named Secunda, and not at Sasolburg, which would increase the Sasol output by a factor of 10 when it reached its full capacity of 58,000 BPD. The plant began production in March 1980, and has not experienced the Sasol I shakedown trauma. The first Synthol unit "achieved all product specifications during its first production run."

Sasol I's history—even after the technical bugs were hammered out—offers other warnings. Although Sasol I and Sasolburg are today, by most accounts, quite clean, they began as an environmental disaster area. For most of the first twenty years of its life, Sasolburg existed "under a shroud of smoke and dirt," and one portion of the plant was labelled by a national newspaper "among the heaviest sources of pollution in any South African industry."[80]

Particularly offensive was a violent odor known as "the Abrahamrust smell," caused by the production of five times as much organic acid as expected. Other sources of pollution were dust from the high-pressure ash-disposal chutes, catalyst dust from the Arge units, flare smoke from the naptha cracking facility, venting of ammonia and nitric acid at the ammonia plant, dangerous gas emissions from the Rectisol unit, smoke and entrained soda ash from the tar acids area and hydrogen sulfide from the smoke stacks.[81] By the early 1970s, the degree of alarm caused by this environmental degradation occasioned "South Africa's most dramatic fight against industrial pollution," a fight which seems to have been effective.[82]

There has also been concern over occupational safety and health at Sasol I. The plant has suffered three serious explosions and fires, as well as deaths by gas poisoning. A tremendous fire erupted in June 1966, after a valve failed during maintenance of the Synthol unit and leaking oil caught fire. Before this could be controlled though, gas lines ruptured and:[83]

> there was an intense gas fire fed by the broken flare line, the fuel gas line, and other sources which could not be isolated. All gas production and consuming plants in the factory were at once shut down. The gas fire could not be put out because of an explosion and poison hazard created by the gas itself, and it was finally extinguished only after pressure had been released on all gas systems.

Sasol I had no sooner recovered from this fire, than in December another blaze consumed the Synthol products recovery area after a compressor ruptured and gas ignited leading to an intense fire which raged until the system depressurized itself. Nine years later, the most serious of all: "an explosion in the gas reforming section killed seven people and caused excessive [sic] damage to the plant."[84] As one Sasol official, probably Rousseau, said, the wild beast turned purring cat "still scratches now and again when it is not circumspectly handled."[85]

Off-gases from the Rectisol plant produce an overpowering stench of hydrogen sulfide around the gasification section, even after the environmental cleanup earlier in the decade. Each year, Sasol sees about four cases of carbon monoxide poisoning, although, none have ever been fatal. One worker was killed some 10 years ago by nitrogen asphyxiation. [86]

On a more routine level, NIOSH investigators have reported that "burns are common," whether from steam, tar or heat, and "hearing loss has been recorded."[87] The tar separation area is said to be very dirty, and three basal cell carcinomas have been reported. As Sasol does not maintain records on cancer—and only minimal records of any kind at all on its Black employees who work the dirtiest and most dangerous jobs—the full extent of such risks is unknown.[88]

3

The Technology of Synthetic Fuels Production

Perhaps the most difficult aspect of dealing with a proposed synthetic fuel plant is the technical jargon. Discussions of synfuel plants inevitably degenerate into talk about such esoteric subjects as *fluidized bed, hydrotreating, Rectisol process,* and *and methanation.* And, almost every book on the subject is filled with diagrams and flow charts that make even the back of your television set look simple by comparison. It is easy to be intimidated into believing that only a Ph.D engineer can make a meaningful contribution to the synfuel debate.

Fortunately, however, you do not have to have a detailed understanding of how a synfuel plant works in order to ask intelligent questions concerning its operation, costs, and environmental impacts. Few citizens who have affected the construction of nuclear power plants, petroleum refineries, or coal fired utility plants have had a detailed understanding of the inner workings of those plants. They simply learned enough about those technologies to talk with experts and to ask intelligent questions.

It is the purpose of this chapter to introduce you to the language of the synfuel industry; to provide you with enough information on how synfuel plants work so that you can feel comfortable learning more. This chapter will have served its purpose if, after reading it, you can talk to company executives who are building a synfuel plant in your community without being intimidated by their technical jargon. To help you refresh your memory and to ease the pain of wading through this material, we have included a glossary of technical terms at the back of this book.

Do not expect easy answers in this chapter because there are none. This chapter does not attempt to tell you which synfuel technology is the best, the cheapest, or the least environmentally damaging. The fact is that even the most expert of experts can not begin to answer these questions. There have not yet been any commercial scale synfuel facilities built in this country and there are over a hundred different technologies at some stage of development in the laboratory or on the drawing boards of various companies. It will probably take years or even decades of work before we see firm answers to these questions. In the meantime, companies, government agencies, and citizens such as yourself must examine proposed new plants one-by-one, making intelligent guesses concerning their efficiency, cost, and environmental impacts.

Synthetic fuels. The term conjures up visions of a large DuPont plant magically mixing various chemicals together to produce oil or natural gas, much like modern petrochemical plants produce nylon or teflon. But the fact is that synthetic fuels are not really synthetic at all. Rather, they are fuels produced from other fuels. A better term for synthetic fuel plants might well be *energy conversion plants,* since what these plants do is take large amounts of coal or oil shale and convert them into useful liquid or gaseous fuels. This conversion process not only produces useful fuels but consumes considerable quantities in the process.[1] For all this effort, only about one to three barrels of oil or 10,000 cubic feet of high Btu gas can be made from each ton of coal and only about two-thirds of a barrel or less of oil can be produced from a ton of oil shale.[2]

Some people consider that the term "synthetic fuels" also includes biomass fuels (e.g. ethanol from grain and methanol from wood) and fuels produced from urban waste. Most also include liquids produced from tar sands as a synfuel. However, in this book we limit our discussion to synthetic fuels produced from coal and oil shale.

Coal is composed primarily of carbon and small amounts of hydrogen. By contrast, liquid and gaseous fuels such as methane and gasoline contain relatively large amounts of hydrogen. In its simplest terms, the production of synthetic fuels from coal involves increasing the amount of hydrogen in the coal. The more hydrogen that is added to the coal, the higher the grade of fuel produced. If relatively little hydrogen is added, a low grade solid fuel is produced. If a bit more is added the result is a fuel equivalent to residual oil or crude oil. If still more hydrogen is added a product equivalent to LPG or naphtha is produced. Finally, if even more hydrogen is added the result is a gas similar to natural gas or methane.

There are many different processes by which the ratio of hydrogen to carbon in coal can be increased.[3] Most coal based synthetic fuel plants produce both some liquids and some gaseous fuels, although for the sake of convenience the synfuel technologies are generally classified by their primary output into one of two major categories—coal liquefaction or coal gasification. Coal liquefaction technologies are, in turn, divided into two major categories—direct and indirect liquefaction.

COAL GASIFICATION

The basic chemical process for producing gas from coal is to react the coal with water. This usually is done by crushing and drying the coal and feeding it into a large vessel, referred to as a reactor or gasifier. Part of the coal is burned to produce heat[4] and the remaining coal is mixed with steam and either air or oxygen in the gasifier. As the temperature of the gasifier rises (from 1500 to 3500 degrees F depending on the process), the coal begins to break down and the hydrogen atoms from the steam combine with the carbon atoms in the coal to produce a mixture of gases.[5]

The *synthesis gas* produced by the gasification of coal is composed primarily of carbon monoxide, hydrogen, carbon dioxide, and water vapor, along with varying amounts of methane. The precise mixture of the gases produced depends on the temperature and amount of oxygen present. If air is used in the gasifier, the gases contain a high percentage of nitrogen and have a heating value of only about one-tenth that of natural gas, or about 100 to 200 Btus per cubic foot. As a result, they are usually called *low-Btu gas*

If oxygen is injected into the gasifier rather than air, less nitrogen is mixed in the gases to dilute their heating value and the result is a product with about one-third the heating value of natural gas or about 250 to 400 Btus per cubic foot. This fuel is generally referred to as *medium-BTU gas.*[6] Low or medium Btu gas can be further processed to produce a gas with a heating value of about 1000 Btu per cubic foot—or *high-Btu gas.*

Regardless of whether low-or medium-Btu gas is produced, the synthesis gas must be cleaned before it is shipped to the consumer or used to manufacture a high-Btu gas. Cleanup is especially important if high-Btu gas is to be manufactured since the catalysts[7] used in the high-Btu process tend to be "poisoned" by impurities in the gas, especially sulfur. Cleanup is also important, of course, to reduce emissions to the environment when the gas is ultimately burned.

Because of their low heating values, neither low- nor medium-Btu gas can be transported economically over long distances. They can be used, however, as sources of energy or chemical feedstocks for utility or industrial plants. In order to produce a high-Btu gas suitable for long distance pipeline shipment, it is necessary to process medium-Btu gas further. This takes place in two steps. First, the ratio of the hydrogen and carbon monoxide in the medium-Btu gas must be adjusted. This step which is usually referred to as *shift conversion* or *water-gas shift* sets up the proper ratio for the next step which is called *methanation.*[8] Second, in methanation, the hydrogen and carbon monoxide are passed over a special nickel-based alloy. This alloy acts as a catalyst and converts the gases to methane and water. After purification and dehydration, the methane can be used as a substitute for natural gas.

Coal Gasifier Bed Types

Coal gasification processes are generally classified into three major types, depending on the type of bed used in the reactor: *fixed* (or moving) bed, *fluidized* bed, and *entrained* bed.

In *fixed-bed gasifiers*, the coal is usually fed into the top of the gasifier.

The coal is supported at the bottom of the reactor by a grate and the air or oxygen and steam are blown upwards through that grate. The remaining ash or slag is drawn off through the bottom of the reactor, and the hot gases exit through the top or side. The major advantages of the fixed bed gasifiers are that they are simple to operate, use a well developed technology, and offer excellent heat recovery. On the other hand, they require coal that has been ground to a specific size and are unable to handle *caking coals*—coals that tend to swell and stick together when they are heated and clog the gasifier.[9] Some of the most common fixed bed gasifier systems are Lurgi, Wellman-Galusha, and Chapman.

In *fluidized bed gasifiers,* finely sized coal particles are suspended by a fairly high velocity (¾ to 10 feet per second) upward flow of gases. The coal particles tend to exhibit liquid-like properties, almost as though they were boiling. This turbulence exposes much more of the surface of the coal to the steam and air (or oxygen) in the reactor thereby increasing the speed with which the gasifier can produce gas. On the other hand, fluidized bed reactors tend to produce a lower Btu gas and have a lower thermal efficiency than the fixed bed reactors. Some of the synfuel technologies employing fluidized bed gasifiers are the Winkler, Synthane, COGAS, U-Gas, and CO_2 Acceptor processes.

In *entrained bed gasifiers,* fine coal particles are blown into the gas stream before it is injected into the reactor. The chemical reactions take place while the coal is suspended in that gas stream. The coal particles are then filtered from the gas and recycled. One of the major advantages of the entrained bed gasifier is that it can use both caking and non-caking coals. The disadvantages of the system include the fact that it consumes relatively large amounts of oxygen, has an unreliable coal feeding system, and a more complicated product recovery system. Examples of some common types of processes employing entrained bed gasifiers are Koppers-Totzek, Texaco, Bi-GAS, and Babcock and Wilcox.

Environmental Concerns: Coal Gasification

If coal were merely composed of carbon and hydrogen, coal gasification would present relatively few environmental problems. The coal could be converted almost entirely to methane and water. However, coal contains, among other contaminants, significant amounts of nitrogen and sulfur. As the result of the high temperatures and pressures found in the gasifier, these elements react to form such significant pollutants as ammonia, hydrogen cyanide, hydrogen sulfide, and carbonyl sulfide. Coal also contains significant quantities of toxic trace metals such as arsenic, cadmium, and selenium. These substances must be treated or recovered as byproducts if synfuel plants are to be operated in an environmentally acceptable manner.

In addition, coal typically contains up to 15% (by weight) of mineral matter. This means that gasifiers produce large amounts of solid waste (mostly slag and ash) that must be disposed of. Since those solid wastes can contain toxic materials they must be disposed of with care.

There are several key points at which the toxic materials in coal can escape into the environment from a coal conversion plant. The precise nature of the problem will, of course, depend on the technology being used. To illustrate the potential points of environmental concern, it is useful to analyze the operations of a Lurgi based high Btu coal gasification system. There are roughly 11 major steps in such a system. They are discussed in some detail here, because many of the steps are illustrative of processes that appear not only in coal gasification systems but also in liquefaction plants and oil shale processes.[10]

1. *Coal Handling and Preparation.* Coal is delivered to the plant where it is dried (if necessary), mechanically crushed, sized (using large screens), and then cleaned to remove the coal fines that might hinder the operation of the gasifier or present an explosion or dust hazard. At this stage, coal dust and the flue gas from coal drying present the major environmental hazards.
2. *Coal-lockhopper.* The coal is then fed to a device known as a *lockhopper* that permits the introduction of the coal into the pressurized gasifier without depressurizing it. The lockhopper works in the following manner: coal is placed into the lockhopper, the lockhopper is closed, pressurized to the gasifier pressure, opened to dump the coal into the reactor, closed, depressurized, and opened to the outside to permit coal to be once again placed inside it. The entire cycle takes about 10 to 30 minutes. The lockhopper is of some environmental concern, since gas from the reactor can escape into the environment if the lockhopper is not properly designed or operated.

3. *Gasification.* As discussed in some detail above, steam and air or oxygen are mixed with the coal in the gasifier under high temperature and pressure. There are a number of possible sources of environmental contamination in this phase. Because of the high pressures and temperatures, volatile toxic materials can leak out of valves, pumps and pipe joints (these leaks are usually referred to as *fugitive emissions*). For technical reasons, toxic gases may have to be vented to the atmosphere when the plant is started up or shut down. Or, in emergencies (referred to by the engineers as *upsets*) the entire contents of the gasifier might have to be rapidly released to prevent an explosion. It is obviously important that these gases be disposed of in an environmentally responsible fashion. In most gasifiers, these gases are vented to a flare stack where they are partially or completely burned.
4. *Ash Removal.* The ash in the gasifier is continuously collected by a rotating grate inside the gasifier and deposited in a lockhopper at the bottom of the unit. This lockhopper operates on the same principle as the coal feeding lockhopper discussed above and can release particulates and toxic trace elements to the environment.
5. *Quenching.* The hot raw gas coming out of the gasifier is then quenched (cooled) and scrubbed to remove any tar, oil, or fine coal particles (coal fines). The material which condenses as the gas cools is called *gas liquor.* The gas liquor is cooled and recycled to be used to quench the raw hot gas. Waste water from the process must be treated to remove various pollutants.
6. *Shift Conversion.* The quenched gases from the gasifier are then passed through a series of reactors that use catalysts to convert (*shift*) carbon monoxide and steam into hydrogen and carbon dioxide. As discussed above, the purpose of this step is to adjust the ratio of hydrogen to carbon monoxide in the gas so as to permit later *methanation* of the gas. One of the main potential dangers from this step involves the possibility of leaks of high pressure gas that is both toxic and flammable.
7. *Gas Cooling.* In order to permit the gas to be purified, it must first be cooled using air and water to about 86 degrees F.

8. *Gas Purification (Acid Gas Removal).* Before the cooled gas can be upgraded to high-Btu gas in the methanation step all sulfur must be removed to prevent poisoning of the catalysts in the methanator. At the same time, carbon dioxide must be separated out. A variety of processes are available to accomplish this purification. One of the more common is the *Rectisol* process. The sulfur and carbon dioxide free gas leaving the Rectisol or other gas purification system is referred to as *sweet gas.*

   The sweet gas at this point in the process is a medium Btu gas often referred to as *synthesis gas.* The synthesis gas can be further processed to produce various liquid fuels or petroleum products (see the section below on indirect coal liquefaction).
9. *Methanation.* The sweet gas (which is composed primarily of carbon monoxide and hydrogen) is then reacted together with the help of catalysts to produce methane. In some plants, the methanated gas is then returned to the Rectisol unit for final water and carbon dioxide removal. The dried, cleaned gas is then compressed before being placed in a pipeline for shipment to the consumer.
10. *Sulfur Removal Processes.* The Rectisol acid gas removal system discussed above produces a number of hydrogen-sulfide (H_2S) laden gas streams from which sulfur must be removed. Among the techniques used to accomplish this are the Claus and Stretford processes. The Claus unit is used on gases with high concentrations of sulfur and the Stretford on those that are sulfur-lean. In some plants the gas from the Claus unit (*off gas*) which still contains some sulfur is then run through a Stretford unit.

    The gases from the Claus and Stretford sulfur recovery units contain both hydrocarbons and sulfur compounds. These so-called *tail gases* must be further treated to obtain an acceptable level of pollutants. Two well-known methods of doing so are the Beavon and Scot processes.
11. *Waste water treatment.* At various points in a gasification plant, liquids condense from the gas stream and must be sent to waste water treatment units to be cleaned up. Tars, ammonia, hydrogen sulfide, phenols, and trace elements are just a few of the substances that must be removed.

COAL LIQUEFACTION

As with coal gasification, the process of coal liquefaction is one of adjusting the carbon to hydrogen ratio in the coal. As noted above, the ratio of carbon to hydrogen in bituminous coal is about 16:1 whereas the ratio in fuel oil is roughly 6:1. The ratio in methane is 3:1. There are three basic methods for changing coal into a liquid fuel. First, the coal can be heated in the absence of oxygen until its breaks down into two major fractions: one that is composed primarily of carbon (called *char*) and the other that is rich in hydrogen and, therefore, is in liquid form. Second, the coal can be broken 'apart with' steam and oxygen to form a gas and then the gaseous molecules can be rebuilt into liquids. Third, hydrogen can be directly added to the coal thereby increasing the hydrogen to carbon ratio and liquifying it.

The first method is referred to as *pyrolysis.* Although pyrolysis is a simple process and can be used at relatively low temperatures and pressures, it produces less than one barrel of liquid per ton of coal. As a result, most of the interest in coal liquefaction has centered on the other two methods which are commonly referred to as *indirect* and *direct* liquefaction. In this section, we will briefly outline the approaches for directly or indirectly liquefying coal.

First, however, it is useful to note that all of the coal liquefaction technologies—whether direct or indirect—tend to have common processing steps, including:[11]

1. *Heating*—all of the processes tend to operate at elevated temperatures and pressures.
2. *Gas-liquid separation*—all of the processes cool and depressurize the materials produced in the main reactor or dissolver. This separates them into liquids and gases.
3. *Acid gas and sulfur removal*—the gas steams in all of the processes are treated to remove acid gases, sulfur compounds, and sometimes carbon dioxide. A number of the technologies employed to accomplish this are discussed above in the section on coal gasification.
4. *Liquid product separation*—all of the processes separate out the liquid products and sometimes distill them into various fractions (e.g. naptha, fuel oil, and diesel oil).

5. *Hydrotreating*—many of the processes will employ a method known as *hydrotreating* to remove sulfur and other objectionable contaminants from the liquid fuel ultimately produced.[12]

Indirect Coal Liquefaction

The indirect coal liquefaction process begins with the gasification of coal. However, after the gas has been subjected to shift conversion and purification it is sent to a *catalytic synthesis unit* rather than to a methanator. The catalytic synthesis unit rebuilds the chemically simpler gas molecules into more complex liquids. Depending on the type of synthesis unit the gas can be built into gasoline (using Mobil's M-Gas process), methanol, or a mixture of various hydrocarbons (using the Fischer-Tropsch process).

Methanol Production. Methanol can be produced from synthesis gas by passing the gas over a zinc or copper catalyst. This catalyst speeds a chemical reaction for recombining the hydrogen and carbon monoxide into methanol without actually entering into the reaction itself.

For many years, the production of methanol was somewhat difficult, requiring a high-pressure step operating at about 8000 pounds per square inch. However, recently Imperial Chemical Industries (ICI) has developed a low-pressure process. Other companies have followed suit and developed their own low-pressure techniques and now there are at least four low-pressure methods available on the market.

Methanol can be used in a variety of ways: as fuel in gas turbines for generating electricity, as feedstock in chemical plants, or as transportation fuel. As noted below, it can also be converted into gasoline using the M-Gas process.

M-Gasoline Process. Methanol has a number of problems when used as a substitute for gasoline in cars, buses or trucks. First, it has only half the energy of gas thereby requiring the vehicles to carry more fuel. In addition, if gasoline and methanol are mixed and they come into contact with any water, the methanol settles to the bottom of the fuel tank. This has led a number of companies to look for a technology for converting methanol into gasoline.

One such process has been developed by Mobil and is known as the M-Gasoline process. Methanol is passed over a series of *zeolite* catalysts which reform the carbon and hydrogen in the methanol into a product similar to high-octane unleaded gasoline. While promising, the Mobil process has not yet been tested on a large scale. A commercial scale plant is under construction in New Zealand.

Fischer-Tropsch Process. The best known and most widely used indirect liquefaction process is the Fischer-Tropsch (F-T) process. This was one of the two processes used by the Germans to produce military fuels from coal during World War II and is the process employed by the Sasol plant in South Africa, the only commercial scale coal liquefaction plant in the world. The F-T process uses an iron or cobalt catalyst to reform the hydrogen and carbon monoxide present in synthesis gas into a mixture of useful hydrocarbons including gasoline, methanol, kerosene, diesel and fuel oil, waxes, and a variety of chemical products.

Although F-T has been commercially proven and produces a wide variety of useful products, it does have significant drawbacks. It has relatively low efficiency and produces less than two barrels of product per ton of coal. This may make the process more expensive than other liquefaction technologies.

Direct Coal Liquefaction

The theoretical advantage of direct coal liquefaction is its higher efficiency. Whereas in indirect coal liquefaction the coal is broken down and then rebuilt to increase the ratio of hydrogen to carbon, in direct liquefaction hydrogen is added directly to the coal. This results in yields of up to 3 barrels of product per ton of coal. On the other hand, direct liquefaction produces raw fuel that is partially contaminated with sulfur and nitrogen, thereby presenting potentially significant environmental problems.

There are two basic approaches to direct coal liquefaction—solvent extraction and catalytic liquefaction (also known as *hydroliquefaction*). Both processes require the use of high pressures—between 500 and 4,000 pounds per square inch.

Solvent Extraction. In this process hydrogen is "added" to the coal using a solvent that is made in the process. This solvent both dissolves the coal and "donates" hydrogen to it. In some of the processes, this *donor solvent* is then treated with a catalyst to restore its hydrogen content and recycled. Some major solvent extraction processes include SRC-II, Consol, Exxon Donor Solvent, and the UOP Process.

To provide a rough idea of how the process works, we will briefly examine the Exxon Donor Solvent (EDS) process. In the EDS process, a solvent is enriched with hydrogen in the presence of a catalyst and then mixed with crushed coal. The solvent and coal mixture is passed through a *liquefaction reactor* along with gaseous hydrogen. Under high temperature and pressure (about 1,500 to 2,100 pounds per square inch and 800 to 900 degrees F), the coal reacts to form a mixture of solid and liquid products. This mixture is then sent to a *flash distillation unit* where the reactor products are separated into gases, solvent, liquids and a solid product called vacuum bottom slurry. The solvent is recycled. The bottom slurry is further treated by coking or gasification to produce additional liquid fuel and hydrogen. The liquids from the process are sent to a *product hydrogenation* unit where they are treated much like the similar fractions that emerge from conventional petroleum refining units.

Catalytic hydrogenation. In the catalytic hydrogenation process, coal is directly reacted with hydrogen in the presence of a catalyst and a solvent. Some major catalytic hydrogenation processes include H-Coal, Bergius (the major process used by the Germans during World War II), Synthoil, and the Dow Chemical process. To provide an example of how this technology works, we will briefly examine the operation of the H-Coal process.

In the H-Coal process, the coal is first dried and then pulverized. It is then mixed into a slurry with recycled oil from the process and hydrogen (produced in a separate coal gasification unit) is added to the slurry. The slurry is preheated and pumped into a reactor (at about 840 degrees F and 2,200 to 3,000 pounds per square inch) where it passes up through the catalyst. The coal is partially dissolved and combines with the hydrogen. The products from the reactor are sent to a "flash separator." In that device they are separated into two major fractions: gases, and liquids and solids. The gas products are treated to remove sulfur compounds, ammonia and other "contaminants." The solid/liquid fraction is then separated into solids and liquids and the liquids are distilled into fuel oil and naptha.

STATUS OF COAL CONVERSION TECHNOLOGIES

There has never been a large-scale commercial synthetic fuel industry in this country. Although low- and medium-Btu gasifiers are being employed in a limited number of industrial plants, no commercial-scale high Btu gasification facilities or liquefaction plants have been built in the United States. Common sense tells you that untested technologies will tend to have more problems, a notion that has been borne out by a number of studies. For example, one recent study examined 43 pioneer process plants built in the chemical, oil, and mineral industries over the past 15 years. It found that a significant fraction of those plants operated at below 60 percent of their capacity despite the fact that it had been expected that the plants would operate at 90 percent of their rated capacity.[13] During the construction of pioneer plants several types of difficulties are commonly encountered including: poor performance of equipment, major equipment failure, fires and explosions, excessive corrosion and erosion, and spare parts and maintenance problems.[14] There is no reason to expect that the pioneer plants in the synfuel industry will be any luckier.

Although there have been numerous small-scale tests of the coal based synfuel technologies, these are of limited value. As the staff of the Synthetic Fuels Corporation recently noted: "promising small-scale tests are not reliable predictors of commercial performance, and take many years to prove out."[15] In short, it will be many years before we know with any degree of certainty the technical, economic or environmental viability of the various synfuel processes.

OIL SHALE

A recent report on the status of the Colorado oil shale industry provides the following summary of oil shale technology:[16]

> Unlike the black gold gushers of the Texas plains and the Middle East deserts, the "oil" under the Colorado sagebrush is not waiting to rush to the surface, or even to flow slowly. With rubbery toughness, it sticks to the rock around it. Chemicals won't dissolve it; crushing won't free it; water won't wash it.

> The only way that has been found to move kerogen out of the rock is to heat it—at temperatures twice as hot as a home oven—until a reaction known as pyrolysis occurs and the rubbery kerogen separates from the rock. This process is known as retorting. The shale, in other words, must be cooked, and for centuries mankind has been trying to learn the best recipe.

In short, in contrast to the complexity of coal-based synfuel technology, oil shale technology is relatively simple.[17] It merely involves heating the hydrocarbon *kerogen* which is found in the shale until it breaks down into a liquid fuel. This requires a temperature of over 900 degrees F.

But just because oil shale technology is relatively simple, does not mean that it is free of problems. The fact that oil shale contains such a high mineral content makes its processing very challenging. The mineral content of oil shale is roughly 80 to 90% by comparison with coal whose mineral content is only about 10%. This means that enormous quantities of oil shale must be handled to recover a relatively small quantity of fuel. To be more precise, one to three barrels of oil can be derived from a ton of coal whereas even the richest oil shale yields only about two-thirds of a barrel of oil. Moreover, oil shale tends to contain significant quantities of sulfur, arsenic and nitrogen that must be removed during processing. To make matters even worse, oil shale tends to swell like popcorn as it is processed. As a result, you end up with more spent shale than you started with. Oil shale's other major difficulty is water: oil shale plants will require large quantities of water but will be located in water-short regions.

Oil shale technologies are traditionally divided into two types, depending on where the retorting of the shale takes place—above ground or in place (*in situ*) underground.

Above Ground Retorting

The underground retorting process begins with a conventional surface or underground mine. This is worthy of note because of the immense scale of operations that will be required to feed a commercial-scale oil shale industry. For example, the U.S. Bureau of Mines has estimated that producing 180,000 barrels of oil shale a day would require a surface mine that would eventually be 2 miles long, 1.5 miles wide, and 2,000 feet deep.

Seven years into its operations such a mine would have 20 huge shovels loading more than 100 dump trucks each capable of carrying 150 tons of shale. This is comparable in size to the world's largest iron and copper mines.[18]

In areas where the shale is relatively far underground, mining would take place using traditional underground mining methods. Not as much of the shale can be recovered in this case since about 40% of the deposit must be left in place to support the roof of the mine. As with the surface mines, the underground shale mines would be among the largest in the world; it would take production of about 100,000 tons of rock per day merely to support a single 50,000 barrel per day oil shale retort.[19]

Once the shale has been mined, it is heated in a large vessel (called a retort) and the resulting gases and liquids are collected. Above-ground retorts can generally be divided into two types: those in which the shale is heated by combustion within the retort (direct heating) and those in which heat is pumped into the retort from outside (indirect heating).

Direct Heating. The *Paraho Direct* process is one type of direct heated retort process. Raw crushed oil shale is inserted into the top of a 10-story high vertical kiln. As the shale moves down the kiln, it comes into contact with an upward flow of hot gases radiating from a zone in the kiln where part of the shale is being burned. When the shale reaches about 900 degrees F, the kerogen breaks down creating an oil mist that rises to the top of the kiln and is withdrawn. The retorted shale continues to fall down in the kiln, until it reaches the combustion zone. In that zone, the residue of the spent shale is burned with air and recycled gas. The resulting hot gases rise to heat the next descending batch of raw shale. Eventually, the shale drops farther down in the kiln where it cools and is eventually removed.[20] The Paraho process can also be operated in an indirect heating mode by recycling hot product gas that has been heated in an external furnace.[21]

Indirect Heating. A type of above ground oil shale technology employing indirect heating is the TOSCO II process. In that process, crushed oil shale which has been heated to about 500 degrees F is mixed in a drum with small ceramic balls (about the size of large marbles) that have been heated to 1200 degrees F. The shale temperature eventually reaches 900 degrees F, thereby causing the kerogen to decompose. The resulting oil vapors are collected, cooled, and *fractionated* into naptha, oil, residue, and gases. The gases are recycled and burned to heat the ceramic balls, the spent shale is cooled and disposed of, and the ceramic balls are cleaned and recycled. One major advantage of the indirect heating methods is that they prevent the gases from the pyrolyized shale from being contaminated by or diluted with combustion products and nitrogen.

Underground (In Situ) Retorting

As we have noted, above ground retorting requires that enormous quantities of shale must be mined, transported, processed, and disposed of. To minimize these problems, considerable effort has been devoted to developing technologies for retorting oil shale in place underground.

The first efforts at in situ retorting involved what is now called *true in situ* retorting. In that process, which is most likely to be applied to shale deposits within 1,000 feet of the surface, shale deposits are drilled and fractured by blasting or the use of mechanical devices. The deposit is then simply ignited and allowed to burn[22], and the oil released from the kerogen is collected in wells and pumped to the surface. The true in situ processes have met with considerable problems. The tight deposits in Colorado leave little room for hot gases and oil to circulate underground. So more and more companies have turned away from in situ processes in favor of *modified in situ.*[23]

The modified in situ technologies (MIS) are a hybrid of the two we have already examined. In MIS, 20 to 40% of a shale deposit is mined out using conventional mining methods and the remaining shale is fractured (*rubbled*) through the use of explosive charges. The rubbled shale expands to fill the mined-out void thereby creating a much more permeable deposit. Finally, the rubbled shale is ignited and the resulting oil recovered and pumped to the surface. Perhaps the most important step in the MIS operation is the blasting. The trick is to obtain the right size rubble in order to assure that the deposit burns uniformly. Proper rubbling has proven to be a difficult art.

Although modified in situ technologies solve several of oil shale's major difficulties (they require much less water and relatively little mining and handling of bulky solids) they also create a whole host of new problems. For example, the burned out retorts provide a source of pollutants that might well be leached into aquifers by percolating groundwater. And, the water that is pumped out of operating retorts will be laden with toxic materials.

Upgrading

The crude shale oil from retorting is generally too thick to be piped and too high in nitrogen, sulfur, and various metals such as arsenic to be safely used or refined in a conventional oil refinery. As a result, crude shale oil must be *upgraded* before being transported. Among the treatment techniques likely to be employed in product upgrading are:

1. *Visibreaking*—With this technique, the crude shale oil is also heated to about 900 degrees F for several minutes or seconds. After it is cooled, the oil's *pour point*—the lowest temperature at which it will flow—is reduced from roughly 85 degrees F to 40 degrees F.
2. *Coking*—With this process the crude shale oil is also heated to about 900 degrees F and placed into a vessel where it is permitted to decompose into various gases, oil and a solid (*coke*). This process produces a more pourable, higher grade liquid. However, the process is rather unattractive because it produces relatively little oil and a considerable quantity of low-value coke.
3. *Hydrotreating*—In this technique, the crude shale oil is subjected to catalytic hydrogenation in order to remove its nitrogen and sulfur. The result of this process is a product that can be burned directly as a fuel oil or further refined into other products. Unfortunately, hydrotreating is a relatively expensive process.
4. *Chemical additives*—Another method of reducing the thickness of crude shale oil is to add various chemicals known as *pour point depressants.* Alternatively, crude oil can be made easier to ship by mixing it with conventional crude petroleum.

Refining

The physical and chemical properties of crude shale oil differs from that of many conventional crude oils. However, depending on the upgrading techniques that are employed, the resulting shale oil can be comparable to the best grades of conventional crude oil and can be refined in conventional oil refineries. It should be noted that shale oil is not as good a source of gasoline as it is of jet fuel, diesel fuel, and heating oil.[24]

The Status of Oil Shale Technologies

The major question mark in the development of the oil shale industry is the technology of retorting. There have not yet been any commercial scale retorting operations and the risks of failure seem high. Most of the test plants that have been operated were only one-tenth to one-hundredth the size of a full scale commercial plant and even at that small scale some of those projects faced formidable difficulties. As one recent major study concluded:

> Many important questions about the technologies associated with retorting of oil shale remain to be answered. Regardless of the approach, several more years of study and testing will be needed before the feasibility of an oil shale project can be fully evaluated.[25]

Another study put it somewhat more bluntly:

> With the present technical status of the critical retorting processes, deploying a major oil shale industry would entail appreciable risks of technological and economic failure. Although much R&D has been conducted, and development is proceeding, the total amount of shale oil produced to date is equivalent to only 10 days of production from a single 50,000 bbl/d plant . . .[26]

Conclusion

There is no doubt that high quality petroleum products can be extracted from oil shale and manufactured from coal. However, there are numerous questions that have yet to be answered concerning this infant industry's costs—both economic and environmental. The purpose of the balance of this book is to help you identify and shape those hard questions. Hopefully, this chapter will have provided you with some of the technical language by which to ask them.

4

Environmental Impacts of Coal Conversion

The purpose of the conversion technologies described in Chapter 3 is to produce cleaner, more flexible fuels from coal. But as one report puts it, if the input (coal) is dirty and the output is clean, then "the pollutants must have been left behind."[1] The question is whether the pollutants can be kept isolated and out of harm to man or the natural environment. The answer to that question is largely yes, but there are many potential exceptions.

The pollutants resulting from coal conversion can be released to the air, the water, or the land. To the extent that emissions via one pathway must be reduced, other waste streams increase. Some of the by-products of coal conversion are similar to those of coal-fired powerplants, coke ovens, and chemical plants. Various environmental laws, discussed in more detail in Chapter 7, require a plant operator to control emissions of certain pollutants also emitted by these conventional facilities. The most advanced and effective emissions control technologies in existence are available for these pollutants. Extensive human health studies have been undertaken to determine the effects of these regulated substances—how they are generated, safe levels of exposure, and how they can be controlled. Yet, not all of the pollutants produced by synfuel plants are regulated. Existing health and safety standards are inadequate to protect workers and the public from these types of pollutants. Moreover, there is concern that "fugitive emissions" or pollutants that escape accidentally or at unregulated points within the facility, may exceed permissible levels.

The most serious environmental impact associated with coal conversion is likely to result from disposal of solid wastes, which will require large areas of land and could ultimately infiltrate and poison water supplies. Potential water pollution from wastewater discharges represents the next most serious threat. While most project sponsors claim that only fully treated water will be returned to the environment, or that there will be "zero discharge" of polluted substances, such a practice may be difficult to achieve in reality. Water consumption may also create an environmental stress in some areas. Air pollution from coal conversion plants poses the least threat, provided that "fugitive emissions" are indeed infrequent.

In a coal conversion plant, pollutants are generated in literally hundreds of places. The rest of this chapter describes some of the most important pollutants, the most likely avenues through which they may be introduced to the environment, and their known health effects. As a first step, we will take a look at the chemistry of coal.

COAL: A CHEMICAL STEW

Coal is a mix of organic chemicals, trace metals, and other substances linked together in a complex, molecular form. Sometimes as much as 25% of coal is composed of chemical "impurities" and metals. Sulfur, oxygen, and nitrogen are the most prominent impurities, but most coals also contain significant amounts of trace metals such as uranium, arsenic, boron, and beryllium.

The percentage of trace metals in coal is quite small, but synfuel plants will process tens of thousands of tons of coal per day, so the total amounts of trace metals generated may be large. One estimate indicates that a proposed coal gasification plant will generate the following trace metals in just one year:[2]

Arsenic— 14,000 tons
Cadmium— 1,000 tons
Lead— 10,000 tons
Manganese—54,000 tons
Mercury— 200 tons

Because some trace elements resist biochemical detoxification and all are nondegradable, major difficulties can arise in disposing of them. Trace elements persist indefinitely in the environment, and have the potential to accumulate in organisms until they reach harmful levels.[3]

When coal is heated or burned in the presence of air, the impurities combine with each other and the chemicals in the atmosphere. Some of the materials produced are quite familiar. The sulfur in coal may form sulfur dioxide (SO_2), one of the most common air pollutants and regulated by the Clean Air Act. Nitrogen from coal will form nitrogen oxides (NO_x), another regulated pollutant. Carbon, after combustion, will be found in carbon dioxide.

But coal combustion (burning) or reduction (heating without oxygen) produces more than these two or three simple chemicals. By one estimate, more than 10,000 substances, only a thousand of which have been satisfactorily identified, are produced by the kinds of reactions which take place in a synfuels plant.[4] Many of these substances may be benign, but a surprising number have been identified as significant hazards to health and the environment.

The bulk of pollutants generated in a synfuels plant emerge ultimately in a solid form. There are both technical and regulatory reasons for plant operators to convert as much potential air or water pollution as possible to solid form. Solid wastes can be managed more cheaply than air and water pollutants dispersed into the general environment. In addition, two characteristics of the regulatory framework encourage the reduction of wastes to solid form. First, solid wastes are not limited by law in amount or by concentration the way air and water pollutants are. Second, solid wastes that are not legally defined as hazardous almost totally escape federal regulation, while air and water regulations apply even when nonhazardous amounts of regulated substances are present.

Not every emission is a health hazard. Many of the emissions in a synfuels plant will come from pollution control devices which are taking a waste substance in a hazardous form and converting it to a less harmful form. For example, one common pollution control process removes gaseous hydrogen sulfide (H_2S), which is a health hazard, and converts it into solid elemental sulfur, which is not. (The hydrogen sulfide itself is a product of another unit that removes acid gases from the main gasification stream. And the unit that removes the H_2S produces its own emission stream that must be cleaned up.)

Potential pollution from synfuels plants falls into three categories: conventional pollutants, trace metals, and organics.

Conventional pollutants are those typically released by direct combustion or burning of oil, natural gas or coal. The conventional pollutants, which consist primarily of sulfur and nitrogen compounds and particulate matter, represent the bulk of pollution that will be emitted from a coal conversion plant. Other important conventional pollutants include the ash and char from burned coal, and sludges from flue gas scrubbing and wastewater treatment. All of these pollutants can be injurious to human health, but are generally required by federal law to be emitted at low enough levels that noticeable injury will not occur.

The *trace metals*, such as mercury, selenium, arsenic, molybdenum, cadmium, beryllium, and flourine, are all poisonous. They may be present in gaseous emissions, water discharges, or solid wastes. Some, such as mercury, arsenic, selenium, and cadmium, volatilize (form a gas) relatively easily; others, like lead and beryllium, are much more likely to be found in liquid and solid wastes.

The *organics* are the most difficult pollutants to categorize. Technically, any compound which contains carbon is an organic. Hydrocarbons are organics that contain hydrogen and carbon. Coal and oil are hydrocarbons. *Aromatic hydrocarbons* contain rings of six carbon atoms. Aromatic hydrocarbons with more than one carbon ring are generally called *polyaromatic hydrocarbons,* or PAHs. Synthetic fuel processes are likely to produce significant amounts of PAHs. Among the most commonly studied PAHs are benzo(a)pyrene, chrysene, dibenzanthracene, and methylbenzacridene. These PAHs are all proven carcinogens. Organics which include chemicals other than hydrogen and carbon are called heterocarbons. Organics called amines contain nitrogen; some polyaromatic amines are also known carcinogens. Thiophenes and mercaptans are organics that contain sulfur. Those that contain oxygen include phenols (an important class of synfuel pollutants) and cresols.

It is not necessary to be able to diagram these complex structures on a blackboard to discuss them at a public hearing or during a permit application. It is important to know that they exist, that they are among the most important pollutants associated with synfuels plants, that they are not yet entirely understood, and that the methods for their control are not altogether dependable. In the next section we describe the steps of the conversion process in which these pollutants are likely to be generated.

SOURCES OF POLLUTION

In addition to the equipment necessary to gasify or liquefy coal, a coal conversion facility may include its own coal-fired steam generator, gasifier, refinery, and chemical plant, as well as dozens of environmental control units. Therefore, it is no surprise that there are many potential sources of pollution in a synfuels plant. The Draft EPA Pollution Control Guidance Document* for Lurgi-Based Indirect Liquefaction Facilities lists no fewer than 114 major waste streams.[5] And these are only the planned waste streams: there are several thousand valves, flanges, seals, pumps, and vents that, if they leak or break down, can be the source of unplanned ("fugitive") emissions.

Pollution may be generated by a conversion plant in three ways: 1) in planned air or water emissions and solid waste disposal, 2) in "fugitive" emissions, and 3) in emissions during accidents. By and large, the planned air and water emissions of a synfuel facility will be less than or comparable to those of a coal-fired power plant. But it is also important to remember that many of the operations in a synfuels facility take place at extreme

*We will refer throughout this chapter to the series of Draft EPA Pollution Control Guidance Documents of which this report is one. Because the Reagan Administration terminated all work on these documents, final versions have not been produced. While the Administration claims the reports are a waste of time and money, we cannot resist the conclusion that it simply is unwilling to acknowledge the results of EPA's own research.

temperatures and pressures and involve materials that can be corrosive or abrasive. The combination of pressure, temperature and corrosivity increases the probability of equipment failure or malfunction. This gives concern about the effectiveness of pollution control technology adapted to synfuels from other industries. Operational problems in a conversion plant may cause the pollution control equipment to break down. For example, a waste stream from the plant may be too erratic to allow the control device to operate effectively. Thus, controls may not be able to maintain emissions at planned levels.

The extreme operating conditions of a synfuels plant also increase the risk of significant fugitive emissions, unplanned escapes of pollutants through valves, flanges, pumps, or other malfunctioning equipment. Unlike planned emissions, fugitive emissions are not released by choice, after appropriate processing and at the point of the plant designers' choosing. They just happen, and some of the areas in a plant most likely to be the site of such emissions have high concentrations of the type of volatile organics which could easily escape into the atmosphere.

A comprehensive analysis of potential fugitive emissions was performed as part of the environmental impact statement (EIS) on one proposed liquefaction plant. It identified more than 5,000 possible fugitive emission sources—intake valves, flanges, pump seals, compressor seals, vents, and open-ended valves. Based on experience garnered from refineries and chemical plants, the EIS estimated that a demonstration plant, less than one-quarter of the size of a commercial plant, might release up to 750 tons of hydrocarbons per year through fugitive emissions.[6] The level of fugitive emissions at a plant will depend upon the reliability of plant equipment and the frequency, consistency, and effectiveness of inspection and maintenance.

Both the point sources of emissions and the fugitive emissions that have been discussed occur under normal operating conditions. A third category of emissions is produced during upset or emergency conditions. Under such conditions, gases that are normally treated or flared to eliminate important pollutants pass directly into the atmosphere. An upset could occur, if a control for a valve or flange was broken or improperly set, or if a deliberate attempt was made to relieve valve pressure to prevent equipment damage or personal hazard. It is difficult to project the impact of an upset in the absence of any operating experience with large synfuel plants, but a feeling for the potential magnitude of such emissions is provided by an estimate made in a Department of Energy EIS on the proposed Solvent Refined Coal-I facility, which indicates that serious upsets at that plant could send as much as 700 tons of hydrocarbon gases to the flare system in a matter of hours.[7]

A final potential pollutant from a coal synfuels plant is the product itself. Conventional crude oil and natural gas are regulated water pollutants, because of the harm they can cause to aquatic life when spilled into streams and lakes. Some coal liquids that have been analyzed contain much higher percentages of toxic materials then conventional petroleum and exhibit far higher levels of mutagenicity.

In the remainder of this chapter, we will examine in more detail the environmental impacts of coal conversion facilities. We will look at the air, water, and solid waste streams associated with various coal conversion processes. This material is designed to track the major regulatory programs which can affect a synfuels plant, by telling you which regulated pollutants are released by synfuel facilities. It is valuable background for using the regulatory permit processes discussed later in this book.

SOLID WASTES

Coal conversion plants will produce large amounts of solid waste—in most cases more than a coal-fired powerplant. Estimates of solid waste generation vary widely. One report suggests that from 20 to 40% of the mass of the feed coal will remain as waste.[8] A survey by the Department of Energy (DOE) indicates that the range may be even greater. Regardless of the exact amount produced, solid waste disposal from coal conversion plants will constitute a major bulk storage problem. The DOE survey estimates that one year's output of ash from a single commercial gasification facility would fill an area 1,000 feet on each side (one million square feet or about 25 acres) to a depth of 20 feet.[9]

Solid wastes fall into three categories: 1) ash and slag, 2) wastewater sludges, and 3) spent catalysts. Ash and slag from auxiliary steam and power boilers are not expected to differ significantly from those produced in coal-fired power plants. But bottom ash and slag from gasification or liquefaction operations may have different characteristics, because they are formed under conditions of higher temperature and pressure.

The ash and slag consist primarily of inorganic oxides, which are relatively harmless in the oxide form. However, these wastes also contain trace elements in concentrations great enough to pose a concern should the wastes be introduced into the water supply, via runoff or "leaching." Trace metals found in a typical gasifier ash include the following: antimony, arsenic, barium, cerium, chromium, cobalt, copper, manganese, rubidium, selenium, strontium, thorium, uranium and vanadium.[10] In a commercial plant, the ash will be quenched (cooled) with process water and will retain some of the water into disposal. No tests have ever been run on quenched ash. If the process waters themselves still contain harmful chemicals, they will, of course, also appear in the ash and any ash leachate.

The second major category of solid wastes includes sludges from air and water cleanup processes. Wastewater sludges will result from flue gas desulfurization ("FGD" or "scrubbing"), biological treatment of wastewater, and raw water clarification (settling). Scrubber sludge will typically be the largest of the three. All of these sludges will contain minute quantities of trace elements. In addition, wastewater sludges from treatment of quench waters will contain phenol and other organics. The Interagency Committee on the Health and Environmental Effects of Energy Technologies found that:[11]

> Bottom sludges from evaporation ponds may contain significant quantities of organics and pose handling and disposal hazards. The formation of toxic gases or the evaporation of harmful materials may also present environmental hazards. Metals may catalyze reactions and micro-organisms may act as biological mediators, further complicating environmental characterization, measurement and monitoring.

The third category of solid wastes is that of spent catalysts. Although the volume of spent catalysts is relatively small, the catalysts include some hazardous chemicals. A variety of catalytic processes is used in many coal conversion operations. No two plants are likely to employ the same mix. The precise chemical make-up of most catalysts is proprietary. We do know, however, the content of some catalysts used in several stages of coal conversion. During the shift conversion step, the preferred catalyst is cobalt molybdate, a highly toxic compound. If methane is produced, a nickel catalyst is used during the methane generation step, and may result in the formation of the toxic and carcinogenic nickel carbonyl. Acid gas removal can use four different materials as catalysts; all four will form wastes with a high metal or organic content and sulfur compounds.[12]

Because of the variety of toxic constituents in all three forms of solid waste, the typical proposed environmental mitigation strategy consists of maintaining the wastes in isolation. The two major options for accomplishing isolation are lined landfills and return-to-mine. In landfill, a clay liner which is presumably impervious is placed below the waste. Return-to-mine, as the name implies, involves using the land which was disturbed by the original mining as the site for solid waste storage.

There is some concern, however, that in either form of disposal isolation may be breached, due to runoff or leaching. In that case, serious contamination of surface or groundwater could occur:

> The potential hazard that coal solid wastes pose to the aquatic environment lies in the relatively large concentrations of accessory elements in the waste and the possibility of acid formation. Accessory elements could be leached from the solid wastes by water in a slag pond or water percolating through a landfill. Pyritic minerals in these solid wastes produce acid when exposed to air and water, and acid could lower the pH of the pond or the pH of the water passing through a landfill. Lowering the pH could increase the leaching of potentially hazardous chemical constituents or directly harm organisms in the affected area.[13]

The study from which the paragraph above is excerpted looked at wastes from 11 coal conversion processes and found that, "Approximately one-half of the leachates generated from the 11 coal solid wastes at their natural pH levels were acutely toxic to young fathead minnow fry."[14] Experimental studies on leaching of fly ash and bottom ash from conventional coal-fired power plants also give rise to concern:

> ". . . the studies showed that selenium, chromium, and boron, and occasionally mercury and barium, were released on simulated leaching, and the concentrations exceeded the values recommended by EPA for public water supplies."[15]

EPA has expressed concern that some leachates could present water quality problems for such uses as irrigation.[15A] Given the damage that can occur if solid wastes do enter water supplies, a well-planned program of disposal site maintenance and monitoring is essential. Such plans should extend to the lifetime of the disposal site, which could be considerably longer than the estimated plant lifetime.

WATER POLLUTION

Coal conversion plants must utilize large amounts of water, which will become heavily laden with pollutants in the process. However, federal law requires that these wastewaters be cleaned before discharge into a river, lake or other water body. Thus direct discharges of plant wastes are unlikely to be a major source of water pollution. To the extent that water pollution does occur, it is more likely to result from leaching or runoff of solid waste piles, as discussed in the previous section, or from spills of the crude synfuel product. Nonetheless, it is important to understand the potential for water pollution from facility wastewater streams in order to participate effectively in the plant permitting process and ensure that the conditions of the permits will result in safe operation of the plant.

When there are water discharges from a plant, they are regulated by the Clean Water Act. Section 402 of the law provides that every point source of water pollution must obtain a permit before it can discharge any pollutant into any receiving body of water. These permits are called NPDES permits (NPDES stands for National Pollutant Discharge Elimination System). NPDES permits are discussed in detail in Chapter 7.

If left untreated, the wastewaters from a synfuels plant could pose serious environmental and health risks. But all of the waste streams will be subject to some sort of treatment—in most cases, at least by bacterial decomposition. Unfortunately, there is no analysis which gives an overall feel for the combined, treated wastewater output of an operating, fully-integrated coal conversion facility. In lieu of such an analysis, we summarize below the major wastewater streams produced in a coal conversion facility.

Wastewater is released from a number of steps in coal conversion, including mining, pretreatment, transport within the conversion facility, gas quenching (cooling), gas scrubbing, gas desulfurization, boiler blowdown, and cooling tower operation. In addition, runoff from coal storage piles and sludge and ash impoundments can be acidic and contain numerous trace metals. Untreated, these wastewaters may contain a variety of toxic substances in potentially dangerous concentrations.

A gasification plant may produce seven types of liquid wastes: 1) particulate scrubber waters, 2) raw gas quench waters, 3) ash quench waters, 4) waste sorbents and reagents, 5) shift condensate, 6) methanation condensate, and 7) miscellaneous wastewaters (general plant clean-up water, storm runoff, accidental discharges, etc.). Not all gasification plants produce all of these waste types.

EPA studies indicate that the particulate scrubber waters represent the largest stream of wastewater, followed by the ash quench waters.[16] No estimate is given for the quantity of shift condensate. An analysis of the Synthane process presented in another study suggests that the shift condensate for that process is approximately equal in volume to the methanation condensate.[17] It should be cautioned, however, that the relative volume of the different wastewater streams will vary among processes and according to coal type. According to EPA, the dirtiest wastewater streams are the waters used to cleanse the product as it first emerges from the gasifier, namely the scrubber and quench waters. The shift condensate and gas purification (acid gas removal) waters are likely to be of intermediate quality. The methanation condensate is expected to be virtually clean.[18]

The gas quenching waters from gasification processes contain high levels of ammonia, sulfide, and thiocyanate, and in the case of the Lurgi process high levels of organic substances. In entrained or fluidized bed gasifiers, the quenching waters will be heavily laden with particulates. Significant levels of trace metals may also be present in the gas quenching waters. For example, at the Sasol Lurgi plant, 90% of the arsenic, 42% of the fluoride, 35% of the cadmium, and 32% of the mercury originally present in the coal were found in the gas quenching waters.[19] Particulate scrubber wastewaters will contain high levels of suspended and dissolved solids, as will ash quench waters, which may also contain arsenic, boron, fluoride, and selenium in significant quantities.

The quality of the scrubber and quench waters, or the process waters, will depend on the type of coal and the gasification process used. The Lurgi and Synthane processes are expected to give the dirtiest water. The Hygas process should give a somewhat cleaner wastewater stream, as should the CO_2 Acceptor process. The process condensate from the Koppers-Totzek gasifier is expected to be "quite clean."[20] As in the case of coal gasification, the largest and dirtiest wastewater stream in coal liquefaction will be the process waters—from sour gas scrubbing, acid condensates and other sources.

These wastewaters will contain phenols, ammonia, and sulfates. In addition, they are likely to contain many of the substances found in gasification waters, including polycyclic aromatic hydrocarbons, chlorides, cyanides, and a variety of trace metals including arsenic, antimony, cadmium, lead, mercury, and nickel.

Because liquefaction processes have greater variation than do gasification processes, it is more difficult to generalize about the sources and quantities of wastewater streams in liquefaction plants. However, the principal wastewater streams in liquefaction processes will be similar to those in gasification. In those liquefaction processes where the coal is first gasified, much of the dirty effluent will come from the gasifier.[21]

Having outlined the quantities and hazardous constituents of wastewaters from coal conversion, we should reiterate here that none of these wastewater streams would be discharged directly into the environment during ordinary operating conditions. In order to obtain a federal water discharge permit, the plant operator must subject all the waste streams to at least a minimal degree of treatment. Most project proposals claim that the anticipated treatment regime will prevent any significant release of pollutants to the environment. This claim can be accurate. However, there are some uncertainties about the ability of conventional pollution abatement technologies to control all the pollutants generated by coal conversion. These uncertainties, combined with the unpredictability of accidental releases, argue strongly for an extensive monitoring program at any conversion facility.

The type of pollution control technology and its expected effectiveness are exemplified in the following statements from the EIS on the SRC-II liquefaction project:

> Process streams and process area runoff will undergo biooxidation treatment to remove most of the phenolic compounds and other organic compounds. All wastewater except cooling water blowdown will receive tertiary treatment consisting of sand filtration and carbon absorption. . . . Following these treatments, a portion of the wastewater will be reclaimed by evaporation or reverse osmosis and recycled as makeup water. . . . The remainder of the treated effluent . . . will be discharged to the Monogahela when river conditions permit.[22]

The EIS finds that this treatment will allow the discharge to meet federal water quality criteria, as long as the river flow is at normal levels. However, during periods of low river flow, wastewater streams would have to be stored or the plant would have to shut down. The report indicates that a discharge of treated wastewater during periods of low flow would result in "borderline levels or noncompliance with standards or criteria for numerous water quality constituents." [23] The report did not analyze, however, the discharge of untreated wastewater.

The report does not give the project an unqualified bill of cleanliness. It finds, for example, that,

> Although the SRC-II discharge is expected to meet water quality standards, there is concern about potential pollutants such as PNA [polynuclear aromatic hydrocarbon compounds, also abbreviated as PAH], for which no standards exist. . . . Although the analysis in Table 4.9 suggests that the SRC-II discharge of PNA will not represent a great threat to water quality, further research and monitoring are needed to determine the impacts.[24]

And it finds that even the discharge of treated wastewater will "constitute a small stress on the aquatic ecosystem."[25] It also concedes that accidental releases can occur, pointing out that,

> Such nonroutine [minor] releases from industrial facilities often account for significant ecological impacts on aquatic ecosystems. . . . It must be assumed that nonroutine releases from the SRC-II demonstration plant will cause stress on the "ecosystem health" of the Monogahela River over and above the stresses caused by routine wastewater discharge.[26]

While the type of water pollution controls discussed above is not new, we should not assume that existing technologies will be entirely applicable to coal conversion plants. The following discussion elaborates on some potential problems that may arise in adapting conventional pollution control equipment to synfuel plants:

> Process condensates from coal conversion bear similarities to process-effluent waters from coking in steel mills, and many initial concepts of water-treatment methodologies for the process condensates have been direct extensions of coking-water treatment technology. *Important differences between the two classes of waters, however, are (1) much larger flows in the case of coal-conversion condensates, and (2) much greater concentrations of dissolved carbon dioxide; for example, 5,000-10,000 ppm vs. 1500-2000 ppm for coking effluent. Other possible differences, not yet established, may be a much higher proportion of polyhydric phenols in some coal-conversion condensates, as well as greater quantities of problem tract-organic compounds, such as PAH's, nitrogen heterocyclics, etc.*[27] (Emphasis added.)

A key feature of the proposed wastewater treatment program for the SRC-II project, and for many other proposed coal conversion facilities, is the assumption of *zero discharge:*

> In the West . . . zero discharge of effluents . . . is generally expected. Synthetic or clay liners and catchment basins in evaporative ponds and solid waste sites will minimize the potential for groundwater contamination by leaching the resulting solids.[28]

Zero discharge means just what the phrase implies: no polluted effluents from the plant will be discharged into a natural water body. It was estimated that the commercial-size SRC-II plant would have to operate under zero discharge for 40% of the time in order to keep discharges within 10% of the river's assimilitative capacity for key pollutants.[29]

Zero discharge is not impossible, but it has not been proven at commercial-scale conversion facilities. It may have benefits, but it does not eliminate pollutants; it simply converts water-borne pollutants to solid waste that must be disposed of. Zero discharge will typically be achieved by evaporation and/or sedimentation ponds. Of course, zero discharge will be maintained only as long as the liner does not develop any breaks or become permeable. Although engineers claim confidently that such basins can be made to remain impermeable for decades, we have no experience with which to corroborate the claim. If the liner does break, groundwater contamination may ensue. Evaporation ponds are also undesirable, because they may result in the dissipation of toxic gases formed from the evaporation products. Metals and microorganisms may catalyze reactions, further complicating the situation.[30]

Even with a full array of sedimentation ponds and evaporators, there are times when a zero discharge system may fail. In some climates, or during certain times of the year, continuous evaporation is not feasible. Heavy rainfall can flood settling ponds and lead to spillover. An extended plant shutdown may deprive the evaporators of their power.

The sponsors of the SRC-II plant proposed for Morgantown, West Virginia, stated both in the draft EIS and in public hearings that the plant would produce no water discharge. Some months later, after comments from the Environmental Protection Agency and environmental groups challenged their assumptions, they acknowledged that zero discharge might not work for the following reasons:[31]

1) lack of operating history on zero discharge units at coal conversion facilities and the risk that unintended emergency discharges might occur frequently;
2) increased energy consumption by evaporators;
3) problems of brine or soluble salt disposal may exceed the problems of adding trace concentrations of salts, metals, or toxics to the rivers; and
4) the difficulty of applying zero discharge in a non-arid Eastern region.

They concluded it would be necessary to design a wastewater treatment process to handle likely discharges.[32] Several other projects continue to progress on the assumption that zero discharge can be achieved.

Wherever plans for zero discharge systems are announced, careful attention should be paid to backup systems and contingency plans for dealing with system failure. When a zero discharge system fails, it may allow untreated, perhaps even concentrated, effluents to flow directly into lakes and rivers.

Another potential source of pollution is the crude synfuel product itself, if it is spilled and reaches ground or surface waters. Oil and grease are conventional water pollutants regulated by the Clean Water Act, because of the damage they can cause to aquatic life when introduced into lakes or rivers. Crude synfuel products—the synthetic oils which are the end result of the process—contain not only the constituents of oil and gas but a variety of other hazardous chemicals. An analysis of the SRC-II liquid shows that it contains much higher concentrations of 30 hazardous compounds than does conventional petroleum. The analysis concluded that "phenols and anilines, which are highly toxic to aquatic life, are present in substantial amounts in coal liquids but are not abundant in petroleum."[33] A second study, prepared by Pacific Northwest Laboratories (PNL) for DOE, indicated that SRC-II liquid was 50 times more toxic than Prudhoe Bay crude (a heavy crude).[34]

These studies focused on the acute toxicity of the SRC-II liquid and not on other long-term effects. Another PNL study has shown that SRC-II liquids are substantially more mutagenic (capable of causing birth defects) than either shale liquids or crude oil.[35] Mutagenicity is generally associated with carcinogenicity. But, as samples of the SRC-II liquid have been available to researchers for only a brief period of time, they have been able only to run Ames assay tests for mutagenicity. The Ames tests are faster and much less expensive than any of the existing tests for carcinogenicity. Because there is a high correlation between mutagenicity and carcinogenicity, the Ames tests are considered a valid test of carcinogenicity.

Concern over product spills is warranted. The tiny SRC-II pilot plant at Fort Lewis, Washington, has had a number of spills.[36] Furthermore, spills in transit from rail and truck accidents or pipeline breaks are possible. These sorts of accidents could deposit a large quantity of synfuel product directly into a major waterway, should they occur. The SRC-II EIS noted that the worst case rail accident (loss of one 20,000-gallon railway tank car at the Harper's Ferry, West Virginia, rail crossing over the Potomac River) would cause a "minor water supply emergency" in Washington, D.C., 50 miles downstream.[37]

Cleanup of synfuel spills may be more difficult than cleanup of conventional oil spills. For example, the synfuel liquids produced by the SRC-II process would sink rather than float in water, so conventional clean-up techniques like skimming and vacuuming would not be effective.[38] Many of the most toxic components of the synfuel product are soluble, so they would be rapidly absorbed into the water and carried along with stream flow. The SRC-II EIS postulates "massive" mortality of aquatic life within 5-15 kilometers (8-25 miles) of a major spill and "selective" mortality as far as 100 kilometers (165 miles) away.[39] In streams smaller than the Potomac River where natural dilution does not reduce concentrations of acutely toxic components, the zone of massive mortality could be more extensive.

WATER DEPLETION

Synfuels plants are likely to use large amounts of water. While coal conversion plants are generally expected to use less water than coal-fired power plants in the same size range, the water requirements of a power plant provide a good idea of the order of magnitude of conversion water requirements. When discussing water needs of energy technologies, it is important to distinguish between "consumption" and "withdrawals." Consumption represents the water that is used up in the process, or water that is not returned to its source. Withdrawal represents the total amount of water taken into the plant from the source. Since some water is typically returned after use, withdrawals are usually greater than consumption.

While the actual consumption depends heavily on the cooling system and on the type of coal used, liquefaction processes typically can be expected to use somewhat less water than gasification processes. The gasification processes are all expected to use about the same quantity of water. The Fischer-Tropsch process is expected to be the most water-intensive of the liquefaction processes. [40]

Most of the water is consumed for cooling and cleaning reactor products. For two of the gasification processes, consumption is split fairly evenly between cooling and process water. In the case of the Lurgi gasification and liquefaction processes, the cooling water is likely to dominate consumption.

Due to reuse, the net consumption is typically only about one-third to one-half of total water withdrawals. [41] Still, while usually less than the water consumption and withdrawals typically associated with a conventional coal-fired power plant, the water requirements of a coal conversion facility will be significant. However, most assessments of water availability for synfuel development are optimistic that the need can be met. Nonetheless, in some cases water requirements may rule out specific sites.

Proponents of synfuels development point to data which shows that even in the arid West, streams have average yearly flows in the millions of acre-feet. (An acre-foot is the amount of water it takes to cover an acre with water one foot deep: it comes to about 326,000 gallons.) Projections by the Water Resources Council indicate that even a 2-million-barrel-per-day regional synfuels industry would consume less than 300,000 acre-feet in total. [42] So, what exactly is all of the concern about water availability?

The concern is real enough. It is not that there is not enough water available in theory—that is, if one could lift up a river basin and tilt it so that all of the water went just where it was needed. But physical and institutional barriers make this impossible. No river known to man delivers a steady, "average" flow, day in and day out. And synfuels plants are not the only projects competing for water. Synfuels plants will obtain the most convenient supplies they can, even if it means displacing other users.

Because of the difference between the legal systems for allocating water, as well as the amount of water available, it is necessary to consider water supply problems in the East and West separately. We will start with the East. It may seem surprising that a region with such large rivers and consistent rainfall could have any water supply problem, but as the discussion below indicates, there may in fact be some water limitations even in the East.

In the East, *riparian* water law prevails.[43] (See footnote for discussion of states with exceptions.) Under riparian doctrine, the owner of land adjacent to water has a right to a reasonable use of that water. In other words, the owner can use a good deal of whatever happens to be available.

Most coal conversion facilities in the East will be located in the Ohio River Basin; some may be placed in the Tennessee or Upper Mississippi Basins. All of the rivers in these basins have generally high stream flow, but like the rivers in the West, they are also subject to drought. Planners usually measure the amount of flow likely in a drought by the historical average flow during the worst week of a 10-year period; this is called the 7-day/10-year low flow.

During periods of low flow, synfuels plants will compete with all other users for scarce resources and will add to the depletion of the already reduced in-stream flows. In-stream flow refers simply to water remaining in a river; that which allows fish to swim, plants to grow, and rivers to be used for recreation and transportation. Even a brief drying up of a river can have catastrophic impacts on fish and wildlife. Lowered stream flow will also leave a river or stream far more vulnerable to the impacts of any water pollutants poured into it, since there is less water to dilute the pollution.

Some experts have proposed that no more than 10% of the 7-day/10-year low flow should be consumed annually by all users in order to prevent water quality problems and interruptions in navigability.[44] By this standard, the water requirements of many coal conversion plants would be prohibitive. For example, the combined water consumption of two closely located synfuel plants (SRC-I and W.R. Grace) and two electric powerplants on Ohio's Green River will use 30% of the low flow.[45] And they are far from the only users.

Comparison of a plant's requirements with the estimated low flow will not provide a complete view of the impact if the estimate does not represent the actual flow at the point where the plant will be sited. Not all plants will be located at the maximum flow point of a river, and some may even be located on smaller tributaries where the average flow is much lower than the amount recorded at the maximum flow point. It will be important to see that adequate flow records are obtained for the exact site of the proposed plant. Permitting authorities should write permits to assure that normal pollution discharges are suspended during periods of low flow.

Some synfuel developers may attempt to avoid the low-flow problem by proposing the construction of reservoirs, allowing storage of water to meet needs during dry years. According to the Water Resources Council, synfuel development on almost every tributary in the Upper Colorado River will require storage and impoundment—dams—to deal with low flow and drought. This solution has its own problems. In addition to destroying river habitats, reservoir construction can increase significantly evaporative losses from a river. If a reservoir is proposed, then calculation of total water requirements must take into account this additional loss to downstream users.[46]

In the West, lack of water is a fact of life and has shaped its water law. Western law deals with the user and diverter of water—the appropriator who puts it to work for society—not the owner of adjacent land. Under the *appropriation system;* first in time is first in right. There are legitimate disputes over how much water legally is available for synfuels in the West, particularly in the Colorado River Basin. Some experts have concluded that current consumption plus projected energy and non-energy water requirements could exceed the average flow by over one million acre-feet per year in the year 2000.[47] Other experts believe that there is enough unallocated water available to support a major synfuel industry.[48]

In most western states, rights to use water are governed by a permit system. The permit indicates the amount that a user may take and also, based on the date of the initial appropriation, the seniority of one's water right. In times of drought, junior appropriators lose out entirely, while a senior appropriator may continue to enjoy every ounce allowed by the permit. In this way, western water law has a measure of certainty that is lacking under riparian law. In times of shortage, western water owners know exactly who has rights to what. (In times of extreme shortage, most western states employ a system of preferences that can override the strict seniority system. Preferences vary from state to state, but they generally rank domestic uses first, then agricultural uses, and finally industrial uses.)[49]

In the Upper Colorado and Upper Missouri Rivers, where most western synfuels facilities would be located, water rights, are governed by a complex system of water permits, court decrees, instream flow requirements, and interstate compacts. Depending on how you count these and how you add in federal and Indian water rights; all of the water in these basins may already be owned. Eighty percent of the existing rights for water withdrawal are for agricultural use.

While water rights are generally similar to property rights, there are some limitations on their sale. In South Dakota, water rights must normally be sold together with the land to which they pertain. In North Dakota, agricultural rights may not be sold for industrial use. Other states limit the amount of water that may be transferred to an industrial use or require legislative approval of transfers. Currently unappropriated water may belong to various states under water compacts, or it may be as yet unsold water in reservoirs maintained by the Bureau of Reclamation. Groundwater (which is actually underground) may also be unappropriated.

A complicating factor which clouds ownership of western waters is the matter of federal and Indian reserved rights. All the land in the West was originally owned by the federal government. While much of it was sold off or given away, the federal government "reserved" large amounts for specific purposes. The Supreme Court held in the famed *Winters* case that, when federal lands are reserved, that reservation carries with it a reservation of water sufficient to fulfill the purpose of the reservation.[50] We generally know these reservations as parks, wildlife areas, and national forests, as well as Indian reservations and military compounds. These rights date from the time of the reservation, which makes them senior to almost all other rights. The federal government, Indian tribes, and the states have been feuding over the nature and extent of reserved rights for decades. At this time, there is no way of calculating exactly how much water may be retained by federal or Indian reservations.

Synfuel plants are newcomers. If they want to enjoy the security that comes with senior rights, they must either buy those rights, or obtain access to unappropriated water. There are several methods for acquiring senior rights: purchase of irrigation rights, pumping underground aquifers, purchase of surplus federal water, and possibly even interbasin transfers.

None of these options is problem-free.

The purchase of irrigation rights could end up diminishing the local agricultural base and changing the character of regions which are now predominantly rural and agricultural. The creation of a high-priced market for water rights could make it more expensive for all farmers who purchase their water, thereby having an effect far beyond those farms which might be shut down, because their water supply has been dedicated to synfuel production.

Pumping underground aquifers could create serious risks. Because our knowledge of aquifer recharge is limited, it is not possible to forecast just who stands to lose when an aquifer is depleted. But already, some of the major aquifers on which agriculture is dependent are being drawn down faster than natural forces can replenish them.[51]

Purchase of surplus water from federal reservoirs could lead to bidding wars for water rights. There is also some uncertainty about the Bureau of Reclamation's ability to market freely large amounts of water from the existing federal reservoirs for industrial purposes.

Finally, interbasin transfers from other water-rich areas, particularly the Columbia River, have been proposed to bring needed water to synfuel development areas. However, even the study of diversions from the Columbia River into the Colorado Basin is banned by federal statute until 1988. And even if interbasin diversions were legally and politically feasible, they would be immensely expensive, involving the construction of a massive network of pipelines or aqueducts, dams, and storage areas.

The adventures of the ETSI (Energy Transportation Systems, Inc.) coal slurry pipeline in obtaining water for its needs provides an example of how a synfuels facility can get access to "new" water. ETSI originally planned to tap the Madison aquifer in Wyoming to meet its water needs and received state approval to do so. After some negotiation, South Dakota, which was not pleased at the prospect of someone pumping an aquifer which supplied South Dakota farmers, offered to sell ETSI water that is stored in the Oahe Reservoir to which the state had rights. ETSI accepted the offer, although "at a price that had South Dakota Governor William Janklow drooling."[52]

In summary, sufficient water can be had, but at a cost. It is the nature of that cost and the question of who eventually pays for this use of water that should be addressed when considering the impact of a synfuels plant on local water supplies. The costs may be economic, social, or environmental, and in many instances may not be borne entirely by the synfuels entrepreneur, but rather will be imposed upon local residents.

Three factors must be taken into account when addressing these questions. First, any water impact analysis must include the effect of the water requirements that will result indirectly from the plant's construction as well as the plant's direct needs. The plant itself represents only a portion of water requirements associated with a synfuels project. Municipal growth resulting from local development, mine use, and reclamation, and the needs of associated industries will all add to water demand and depletion. Second, the environmental cost of water depletion as streams dry up or are diverted to satisfy these needs should be carefully evaluated. Recreational uses, wildlife habitat, and wetland areas may suffer. Third, synfuel water requirements must be weighed against existing interests in and claims over who should have water, and in the context of what financial burden will be placed on the public to make the water available.

AIR POLLUTION

Most of the air pollution from a coal gasification or liquefaction plant will be similar to that emitted from a coal-fired power plant, and will be controlled to levels far less than that which typically occurs at a coal plant. The Lurgi gasification process stands out as the process emitting the most pollutants.

While a synfuels plant, by itself, will probably not put enough of these conventional combustion products into the air to cause a health hazard, a single plant can make enough of a dent in an area's air quality that a second plant, or other industrial development, would cause a violation of national standards. If a single synfuel plant consumes one-half or two-thirds of the "clean air" margin, that means that only the remaining amount is left for all new construction of any type. When the increment is used up, no new facilities may be built.

Although it is expected that air emissions from a coal conversion plant will be relatively small, there is potential for pollution from various process steps, and it is a good idea to know these steps and their potential pollutants.

The first potential potential emission source is coal handling, storage and pretreatment. This portion of the process produces mostly fugitive dust: from coal storage piles, crushing, screening, storage, and feeding.

The second potential emission source is the coal and ash lockhopper gases and other transient gases from the gasifier (called transient waste gases). These gases can contain trace elements, tars, phenols, organics, nitrogen compounds, sulfur compounds, hydrogen cyanide, oil aerosols, miscellaneous particulate matter, and aromatic compounds containing sulfur and nitrogen.[53] Because most of these gases have fuel value, present plans call for them to be collected and burned. If incineration is to reduce their hazardous potential, however, a special incinerator will be required. Another proposed control method is recompression and recycling in the gasifier. This method, however, may not be available at all times during the operation of the plant. Control methods for incineration of lockhopper discharges or analogous operations in other gasifiers and liquefaction processes that employ gasifiers will be set on a case-by-case basis. Given the extremely hazardous nature of this waste stream, it is important to pay close attention to how these gasifier waste gases will be treated.

The other major stream coming out of the gasifier is, of course, the product gas itself. As it emerges from the gasifier, it is not sufficiently pure for use either as the base for methane (in gasification) or synthesis gas (in liquefaction). A number of purification and upgrading streams are, therefore, necessary to remove contaminants. The removed contaminants become components of gaseous, water, and solid waste streams.

After tars and oils have been removed from the product gas, it is necessary to remove H_2S and other sulfur-bearing compounds, to prevent catalyst poisoning and to obtain a gas with the appropriate chemical balance (predominantly CO and H_2) for the subsequent reactions. To do this, almost all proposed synfuel plants will employ the Lurgi-licensed Rectisol process, in which CO_2, H_2S, and other compounds are physically absorbed in cold methanol. This operation is also commonly referred to as an acid gas removal (AGR) process. Untreated Rectisol offgas streams cannot be allowed in the plant or in the ambient air. An array of pollution control devices will be used to treat the Rectisol offgas and remove the most harmful chemicals.

Another major air pollution source in an indirect liquefaction plant will be offgas from the synthesis process. Three synthesis processes are commonly used, Fischer-Tropsch, methanol, and Mobil M-Gas. Each of these processes produces some air emissions, predominantly carbon monoxide and light hydrocarbons. In the M-Gas process, CO in the offgas stream is in concentrations well above the one-hour acute exposure level. Some of the light hydrocarbons are also present in concentrations sufficient to create a fire-hazard.

The catalysts used in the synthesis step, as well as those used in any needed shift conversion step (where additional hydrogen is provided by reacting carbon monoxide with hot steam), need occasional "regeneration," often accomplished by heating them to drive off contaminants that have been absorbed onto their surfaces. Eventually they must be decommissioned. Though intermittent, catalyst regeneration and decommissioning is another important, potential stream of air pollutants.

Heat and steam are necessary for any coal synfuel process, and they are commonly provided by burning either raw coal or a process-derived gas. Steam and power generation flue gases are therefore another key source of emissions. Coal combustion produces large emissions of sulfer dioxide, articulate matter, and nitrogen oxides. In fact, the steam boiler may be the largest single point source of emissions of these three regulated pollutants within the entire synfuels facility. Fuel gas desulfurization equipment like that on a conventional coal-fired plant can control most of these emissions. But these are not the only pollutants associated with direct coal combstion. Trace elements including chromium, beryllium, arsenic, nickel, lead, and barium may also add to the total trace element loading caused by the facility.

If the actual substance being burned is not coal, but rather a process-derived liquid or solid or some mixture of these with coal, then even more careful scrutiny should be given to emissions from the steam plant. There has been little experience with the large-scale burning of these alternatives to coal. If the combustion input contains feeds from such highly polluted streams as the lockhopper gas, there is further reason to be concerned about potential environmental problems. Control technologies that adequately limit unwanted emissions in a coal or oil boiler may be of little value in handling process-derived liquid or gas combustion, particularly if they include a large concentration of organics.

Wherever large-scale combustion and heating take place to make steam or electricity, there is likely to be a need for evaporative cooling towers.

Cooling tower drift consists of the condensed or particulate matter from cooling towers, and may include entrained material. The concentrated minerals in the drift could have an adverse effect on nearby vegetation. The more serious concern about drift, however, is that it could become contaminated by leaks in the cooling system. In water-poor areas, raw gas liquor may be used as makeup water in the cooling system. If this is the case, then significant concentrations of ammonia, hydrogen cyanide, and reactive hydrocarbons may be present in cooling tower drift. EPA estimates that if untreated plant process waters are used for makeup water, expected emissions might reach 6 tons per day for ammonia and more than 100 tons per year of volatile organic hydrocarbons (which would inevitably include some carcinogens).[54]

A final series of areas with direct air emissions are the many storage facilities in a synfuels plant. A full-scale liquefaction plant is likely to have separate storage facilities for methanol, gasoline, diesel oil, heavy oils, ketones, heavy alcohols, tar, naphtha, and phenols.[55] All of these are either products or by-products of most liquefaction processes.

There are specific regulations for petroleum storage vessels of greater than 40,000 gallons capacity.[56] These regulations apply to oils made from coal or shale as well as conventional petroleum. Emissions from naphtha storage, tar storage, and gasoline storage are of most concern. Naphtha storage devices are prone to emit potentially toxic levels of hydrogen cyanide, methyl mercaptan, and ethyl mercaptan, as well as troublesome concentrations of benzene. Tar storage vessels will emit PAHs, including anthracene, benzopyrene, and acridine—all of which are active carcinogens. They will also emit thiols, which may create an odor problem. Gasoline storage tanks may emit some aromatics including benzene and toluene, which have chronic toxic effects. There is almost no experience with the storage of ketones and heavy alcohols, but storage vessels for both are likely to be needed.

Despite the existence of sophisticated air pollution control equipment, there are two avenues by which the gases produced at various stages within the plant can be transmitted to the environment. The first is the so-called *fugitive emissions,* which may include both the conventional combustion products and the more toxic trace elements and gases. The second is an issue already discussed under water pollution, namely, the question of to what extent existing control technologies will prevent emissions of "unconventional" pollutants. As the Congressional Research Service put it,

> Although it is likely that technology can provide a means whereby trace element pollution can be dealt with, the problem has not been sufficiently characterized, especially for larger operations, to ensure that this is the case . . .[57]

Claims that a conversion facility will meet federal air pollution standards apply only to "criteria" pollutants: SO_2, NO_x, hydrocarbons, and particulates. Environmental impact assessments need also to address the question of emissions of trace elements and toxic gases.

It is difficult to predict the level of fugitive air emissions to be expected. The EIS on the proposed SRC II project estimated that fugitive emissions from the demonstration plant would release about 100 tons of hydrocarbons per year, assuming the implementation of a directed maintenance program. Without directed maintenance, 750 tons per year might escape as fugitive emissions from the plant.[58] EPA estimated that the Fischer-Tropsch unit of a full-scale liquefaction plant might emit as much as 1,000 tons of air pollutants by itself.[59] The majority of these emissions are also likely to be hydrocarbons. Hydrocarbons are not a regulated pollutant, although they are used by EPA as a "guide" because of the part they play in the formation of photochemical oxidants and smog. EPA has not dealt with their potential role as carcinogens.

The most easily escaping fugitive emissions will be the lightest, most volatile hydrocarbons. In general, these are less likely to be carcinogenic than the heavier fractions which are less likely to vaporize and escape.[60] But it would be overly optimistic to assume that only the "good" hydrocarbons will emerge through loose valves and flanges, especially when high temperatures and pressures are involved. Under extreme conditions, even the heaviest hydrocarbons may escape.

Our discussion of synfuel-generated air pollutants has centered on the health and environmental impacts of emissions within the plant and in the immediate geographic vicinity of the plant. Synfuel plants also contribute to the general carbon dioxide and acid rain problems created by all fossil fuel combustion. No single synfuel plant will contribute significantly to these problems on a global or regional scale, nor will the entire synfuels industry be a major contributor when compared with all the coal-fired electric powerplants in operation. But the choice of increased reliance on fossil fuels, including synfuels, will clearly exacerbate both problems.

MINING THE COAL

A synfuel plant will use substantially more coal than a coal-fired power plant with the same energy output. It is estimated that a large power plant uses nearly 13,000 tons of coal per day. In contrast, a standard-sized (50,000 b/d) liquefaction plant will use between 20,000 and 30,000 tons per day; a full-fledged (250 mmcfd) coal gasification plant could use up to 38,000 tons per day of lignite.[61] This translates into an annual consumption of between 6 and 14 million tons of coal per plant, or from 120 to 280 million tons of coal during a plant's lifetime.

This is coal mining on an immense scale. Only a handful of mines (eight in 1979) produce more than 6 million tons per year—all of them western strip mines.[62] While there is no requirement that all of the feed for a synfuels plant come from a single mine, the costs of coal transportation almost surely dictate that the coal come from a single large mine or several closely grouped smaller mines and, as is typically the case with coal-fired power plants, that they be located very close to the synfuel plant itself. Almost all of the announced synfuels projects conform to this pattern.

This means that any area selected for a synfuels plant will also have to absorb the impacts of coal mining as well. And the areas targeted for synfuels development already face major environmental impacts from existing and planned coal mining for conventional uses.

Although some eastern synfuels plants may use deep-mined coal, it is anticipated that most plants will rely on strip-mined coal. The extent to which damage results from strip-mining is highly site-specific, but past experience shows that significant degradation must be accepted wherever large-scale strip-mining is undertaken. The effects of strip-mining have been well documented elsewhere and will not be addressed in this manual.

OCCUPATIONAL AND PUBLIC HEALTH IMPACTS

The main health impact of coal conversion facilities will be on workers, rather than on any nearby inhabitants. This section summarizes the potential health effects for each group. A more detailed discussion of the effects of each toxic substance produced by coal conversion can be found in the accompanying Appendix.

Effect on Worker Health. Because no commercial-scale coal conversion plants yet operate in the United States, it is difficult to predict the effects on worker health. However, pilot plants, foreign facilities, and similar processes in other industries have been in use long enough to provide some indications of the types of health effects that may be expected. Workers in direct liquefaction facilities may face a greater risk than indirect liquefaction workers because of the greater concentrations of heavy, complex liquid organic materials in the direct liquefaction process streams and products.[63] Gasification workers also generally face a lesser risk than direct liquefaction workers.

As discussed in the earlier portion of this chapter, the gaseous, liquid, and solid wastes from synfuels plants all contain a wide variety of known carcinogens, irritants, and poisons. Workers are intended to be protected from these substances by federal occupational health and safety standards. But the dangerous substances occasionally may be present in quantities exceeding federal safety standards thousands of times over. Despite precautions, worker exposure to these hazardous materials is not unlikely, due to accidental emissions resulting from the abrasive qualities of input streams and the high pressure and temperature conditions required for the conversion processes. For example, an accident at the Exxon Donor Solvent pilot plant in Baytown, Texas, resulted in the spraying of two workers with liquefaction residues. The accident was caused by a plugged coal residue recycle line which burst.[64]

Leaks in process equipment can also increase the potential for worker dermal (skin) and respiratory contact. Burns and fires may also occur during repair exercises, the frequency of which may be high due to the newness of the conversion processes. Resulting worker exposure can increase a worker's risk of getting cancer or susceptibility to other physical ailments.

The actual danger to workers posed by carcinogens and chemicals depends to a large extent on the length and intensity of their exposure to the material. Exposure will be greatest in liquefaction and gasification technologies that produce liquid by-products. There, the most important potential sources of carcinogens are the condensation of reactor vapors and the subsequent separation of tars from liquids.

Industry maintains, for the most part, that the likelihood of exposure to the vast majority of dangerous substances is extremely low. Project proponents often point out that the earlier facilities for conversion and conventional coal processes included little or no industrial hygiene or careful work practices. Today, most proposed conversion facilities have a detailed worker protection program. The program outlined for the SRC-I plant is typical.[65] Such programs are essential, if worker exposure is to be kept to a minimum.

Because the primary mechanisms for worker exposure to biologically active materials will be dermal and respiratory, the industrial hygiene program should be designed specifically to control these routes. A good industrial hygiene program begins with worker education regarding the nature of the hazardous process materials, the mechanisms of exposure, and the protection procedures to minimize exposure and the importance of using them. Hygienic practices should include the use of protective clothing (overalls, an impervious rain suit, face shield, goggles, hearing protection, special boots, gloves, and a respirator), segregated "clean" and "dirty" areas for clothes changing, and mandatory showers.

Equally important to good hygiene is medical monitoring. Physical examinations should be required before employment and annually thereafter. Early detection of medical problems, should they occur, increases the likelihood of successful treatment. Employee records should be kept for a period of time suitable to the long latency period of cancer, and should include demographic information as well as comprehensive work histories, exposure monitoring data, and medical surveillance results. Additionally, there should be continuous review of worker exposure and health data. Where appropriate, work practices should be modified to respond to experience.

These work practices, if employed, will reduce occupational exposure to hazardous substances. But even a good industrial hygiene program may not be able to maintain the type of total protection that industry claims is feasible. It is rarely possible to design industrial hygiene programs to achieve zero exposure. Since there are no hard scientific data concluding that there exist safe exposure levels for carcinogens—below which no effects will occur—one cannot say that no worker health effects will result. Thus, while a strong worker protection program is desirable, it does not guarantee prevention of the types of health effects discussed below.

Cancer. The organic chemicals produced in coal conversion generally have heavy, complex molecular structures. Though not all such substances have been tested for carcinogenicity, many have been clearly implicated.[66] For example, benzo (a) pyrene, dibenzanthracene, chrysene, methylbenzacridene, and certain aromatic amines have all been shown to cause cancer in animals. Evidence indicates that PAHs may also be mutagenic (cause chromosome damage). Certain primary aromatic amines isolated from coal tar also have been identified as causing human bladder cancer.[67] A recent DOE report states that the heavy distillate fractions of SRC-II materials have caused cancer in animal tests.[68]

Epidemiological studies (the statistical analysis of the incidence of a disease within a sample population) are the most certain proof that a substance causes cancer in humans. One such study performed on synfuel workers in this country involved 359 employees in a coal liquefaction plant in Institute, West Virginia, operated by Union Carbide from 1952 to 1962. The study, conducted from 1955 to 1960, reported 16 to 37 times the expected rate of skin cancers.[69] Deposits of a known carcinogen, benzo(a)pyrene, were observed on the skin of workers exposed to high concentrations of airborne coal-derived oil vapors. The conclusions of this study cannot be accepted without qualifications, due to the small number of workers involved and the lack of control technology and industrial hygiene present at the plant. Nevertheless, it suggests that serious worker health problems may be associated with synfuels production.

There have been no direct human studies on the health effects of gasification, although again there is evidence that the incidence of cancer among coal gasification workers may be higher than among the general population. Epidemiological studies of occupational exposure to hazardous material in processes similar to coal gasification suggest that gas workers, coal hydrogenation workers, and coke oven workers are at especially high cancer risk.[70]

Epidemiological studies require large sample populations in order to be reliable. Because the pilot plants in this country and abroad have been small, there are no large groups of workers on which to base statistically valid studies. Use of the statistical approach is also made difficult by the long latency period associated with the appearance of cancer in a population. Cancer does not usually develop until 15 to 20 years after exposure to carcinogenic substances; most American experience with synthetic fuels has been more recent.[71]

An accepted alternative to epidemiological analysis is animal experimentation. Animal testing has been in use since 1915. Coal tar was the first material shown to induce cancer in lab animals.[72] The tests consist of exposing animals to the suspect substances in high doses over short periods of time. These tests are carefully designed to simulate human low-level exposure over many years. Animals are exposed to the substances by the same routes that workers would be: oral, respiratory, subcutaneous (below the skin), or dermal. Animal testing has already identified many carcinogens produced during conversion processes, especially organic substances such as polycyclic aromatic hydrocarbons, and aromatic amines.[73]

In the absence of epidemiological or animal studies, researchers employ short-term biochemical studies to evaluate the potential carcinogenicity of synfuels products and byproducts. The most commonly used of these is the bacteria test known as the *Ames test.* The Ames test is a rapid assay of a chemical's ability to mutate certain bacteria. There is a high correlation between a chemical's tendency to cause mutations in test bacteria and carcinogenicity.[74] Various substances found in coal conversion processes have been found to be mutagenic in the Ames test, among them the heavy organic fractions of Solvent Refined Coal.[75]

These testing methods have revealed that other substances, besides the heavy organics, are carcinogenic. Certain trace metals can be carcinogenic, including arsenic, beryllium, cadmium, chromium, cobalt, iron, lead, nickel, and selenium. Exposure to trace metals is most likely to be a problem for workers handling the liquid wastes where they are commonly concentrated: ash and gas quenching waters, process waters, raw gas liquor, and liquefaction and product upgrading streams.

Coal conversion technologies may produce co-carcinogens, substances that enhance the carcinogenicity of other substances. A synergistic effect has been discovered between ultraviolet radiation and coal tar. This problem may affect workers in the mountainous West more severely than those in the East, because ultraviolet radiation is more intense at higher elevations.

Chemical Hazards. Cancer and genetic changes are not the only deleterious effects of exposure to coal conversion material. Other hazards that workers may face include a wide range of acute physical reactions to chemicals and materials. Some of these conditions are temporary, while others result in permanent disability. They include asphyxiation, eye and skin irritations, lung disease, central nervous system impairment, liver and kidney malfunction, and reproductive disorders.

Asphyxiation is the most imminent hazard in a coal conversion plant. Hydrogen sulfide and carbon monoxide emissions can result from a leak or a breakdown of normal operations. Both can cause asphyxiation without warning. Coal lockhopper gases contain both of these substances in quantities more than sufficient to cause death. Chronic low-level exposure to asphyxiants can cause systemic disorders due to continual oxygen deficiency in critical organs of the body.

To demonstrate some of the problems that coal conversion plants have had with exposure to asphyxiants in the past, consider the experience of three plants in the United States: those of the Glen-Gery Corporation in Reading and York, Pennsylvania, and the World War II era unit at the Holston Army Ammunition in Kingsport, Tennessee.

Although the Reading gasifier is said to be a model of pollution control and cleanliness, it has had carbon monoxide leaks resulting in concentrations of 500 ppm.[76] For purposes of comparison, the current Federal occupational exposure standard for CO is 50 ppm, and NIOSH wants to lower that to 35 ppm with an absolute ceiling of 200 ppm.[77]

CO leaks from pokeholes were found when the Radian Corporation examined another of Glen-Gery's small Wellman-Galusha low-Btu gasification units at its York, Pennsylvania, plant. Radian concludes that, while dangerous compounds are produced, they can be controlled at "negligible costs." "Some of the controls," however, "have not been adequately demonstrated on coal gasification systems," and the danger of toxic leaks remains present.[78]

Much more dangerous are the aged Chapman (Wilputte) gasifiers at the Army's Holston, Tennessee, ammunition plant.[79] Hazard levels are 100 times greater than for the Glen-Gery units,[80] with "high degree" health hazards posed by two waste streams. Although the gasifiers are old and considered obsolescent, having been installed in 1943 to synthesize chemicals for ammunition and conserve wartime natural gas, the plant continues to be operated.

Eye and skin irritations may also affect coal conversion workers. The eyes and skin are frequent targets of irritation in other industrial processes. Fortunately, most irritant responses are also temporary. Burning, inflammation, discoloration, skin lesions and certain types of acne are common responses to chemical exposure. Some compounds may travel through the skin to affect other organs.

Fatal pulmonary edema, or seepage of fluids into the lung, may result from inhalation of nitrogen dioxide and nitrous oxides.[81] Acute exposure to ammonia, sulfur and nitrogen oxides, carbon disulfide, and hydrogen sulfide may cause laryngitis, bronchitis, or emphysema-like damage.[82] These irritants are found in nearly every gaseous and liquid waste stream in a synfuels facility.

Central nervous system impairment may result from exposure to heavy metals, certain organic solvents, carbon monoxide, nitrogen oxides, carbon disulfide, cresol, and benzene. Most of these compounds are ubiquitous in gaseous and liquid wastes from coal conversion. Symptoms of exposure include increased nervousness, irritability, depression, anxiety, mental confusion, tremors, visual disorders, and sleep disorders. Such problems were seen at DOE's HYGAS plant when leaks from faulty pump seals resulted in worker exposure to the chemical toluene. Toluene losses due to pump seal drippings have been estimated at 84 gallons per day in the 75-ton-per-day plant.[83] Toluene oil is absorbed readily through skin contact or vapor inhalation, and can lead to paresthesias (tingling sensation), and coma.[84]

The liver and kidney collect and dispose of toxins in our bodies. For this reason, they are especially susceptible to damage through chemical exposure. The danger of liver damage is enhanced because the liver is capable of functioning even when severely damaged without any recognizable symptoms.[85] Cresol, carbon disulfide, and coal tar products may cause liver cancer or necrosis and kidney malfunction.

A great deal of uncertainty surrounds the possible reproductive effects of exposure to compounds present in coal conversion facilities. Evidence indicates that cadmium may be mutagenic (causing chromosome damage). Hydrogen sulfide, carbon disulfide, mercury, and cadmium have also been found to cause birth defects.[86] Pregnancy complications and various ovarian and menstrual disorders have been noted in women exposed to carbon disulfide. Exposure to lead may also have reproductive effects.[87]

Effects on Public Health in Local Populations

Although the workers in the synfuels industry will face the greatest health hazards, the public well-being is also at some risk. The population surrounding synthetic fuel plants may be exposed to carcinogens and other toxic substances if the air and water pollution streams from the plants are not controlled all of the time, due to failures of pollution control equipment or fugitive emissions. Although the public will experience less concentrated exposures to the dangerous substances than is likely in the case of the workers,[88] the health effects from chronic low-level contact can be significant. Depending on the severity of accidental emissions of toxic substances and the proximity of the neighboring population, a coal conversion plant does raise the risk of increased cancer incidence.

As discussed earlier, the air and water emissions from synfuel plants are expected to meet federal standards, which are designed to protect public health. Because it is assumed that facilities will always comply with the standards, there has been little analysis of their potential impact on the health of the surrounding population. Some indication can be provided of potential health impacts, however, by looking at the health effects on local residents of heavy industry with similar emissions.

Several studies suggest that living near heavy industry may increase individual risk of cancer. In 1973, a survey of the 30 most industrialized counties in the nation revealed that these counties had a significantly higher number of cancer deaths than less industrialized counties.[89] Another study linked an increase in cancer deaths to counties within the United States that contained petroleum industries.[90] These studies are not conclusive evidence, but they are troubling. Effective monitoring for all dangerous substances is critically important to identify public health hazards early.

The many factors that must be examined in an epidemiological study make it virtually impossible to isolate a single cause of cancer among the general population. A case in point is the ambiguity surrounding the California Department of Health Services's recent discovery that the incidence of lung cancer among white males living in the industrialized areas of a San Francisco Bay area county was 40% higher than among those living in the non-industrialized areas of the same county.[91] Although this result was not observed in white females, the cancer difference among white males was evident even at relatively young ages, indicating that occupational exposure was probably not the cause. A further correlation was found between areas with higher levels of sulfate air pollution and elevated rates of lung cancer among white males. Sulfates are not suspected of causing cancer but may be associated with another substance that does. The department is continuing this study to explain the incidence of lung cancer.

Generally, the people who work in a plant also live in the vicinity of the plant. Therefore, the distinction between occupational and environmental exposures is difficult to establish. Further epidemiological studies of the situation may never clearly isolate a single factor responsible for the increase in cancer.

COAL GASIFICATION OVERSEAS

While technologies may exist to control the waste streams from synfuels plants, they must be installed and maintained to be useful. Because of the lack of operating experience with them, there is concern that emergency emissions and unknown leaks may be dangerously frequent. Some case studies of gasifiers overseas illustrate this point.

Information on foreign gasification plants is available from a few reports and conversations with people who have visited overseas plants. Possibly the worst offending and certainly the best studied gasification plant is the Lurgi facility in the Kosovo region of Yugoslavia. The Kosovo plant—located in the midst of a large mine-mouth industrial complex which also includes a coal-fired powerplant, processing and storage facilities, a fertilizer plant, and a steam generation plant—produces medium-Btu fuel gas and ammonia feedstock, as well as naphtha, medium oil, light tar, and crude phenol. The plant was built in the early 1970s by Lurgi, but has been troubled and "operating with some difficulty"[92] ever since.

Kosovo contains virtually *no* equipment to control pollution or in-plant dangers. Described by one American scientist as an environmental and occupational "worst case," the Kosovo Lurgi plant has an international reputation; *Chemical Engineering* magazine has reported that the Kosovo "has been called 'incredibly dirty' by a West German source, and 'a good example of an uncontrolled plant' by an EPA official."[93]

Recent studies for EPA by the Radian Corporation bear out the harshness of these informal judgments. The plant churns out a wide variety of dangerous pollutants, in large quantities and in all media—gas, air, and water. The plant is a maze of what Radian called "uncontrolled discharge streams," of which at least 70 could result in significant environmental damage—so many, in fact, that Radian could not economically study them all and limited its study first to 50, and then 25, of the worst offenders.[94] "Many," reported Radian, "have the potential significantly to impact the environment."[95] The most dangerous of the Kosovo gas streams—the primary escape routes for a multitude of hazardous chemicals—were an ammonia vent and the flare systems for CO_2 and H_2S-rich waste gases (which worked as if they were designed to vent carcinogenic PAHs), along with vents from storage tanks for by-product naphtha and phenolic water, waste gases from the tar separation unit, and both the low- and high-pressure coal lock vents (which released more PNAs in tar aerosols).[96]

At Kosovo, the most environmentally "significant" (of the uncontrolled pollutants dumped into neighboring air in gas form featured: ammonia, hydrogen sulfide, mercaptans, phenol, benzene, toluene, xylene, hydrogen cyanide, and carbon monoxide.[97] Kosovo wastewaters (from the gasification section and the phenosolvan treatment unit) were highly alkaline ($pH = 9.6$) and included "phenols, cyanides, sulfides, and total organics," some at dangerously high levels. In addition, extremely high trace element levels were found in two pollutant streams—arsenic concentrations exceeded EPA Discharge Multimedia Environmental Goals (DMEGs) 850 times over, mercury was 14 times too high, and too much chromium, nickel, cadmium, and beryllium also was dumped.[98]

Solid wastes included a fairly harmless ash, at least in terms of RCRA leachate standards, but also a heavy tar rich in "highly toxic organic materials, such as phenols and PNAs [PAHs]," including the cancer causing benzo(a)pyrene (BaP) and 7,12-dimethyl benz(a)anthracene. The ash is piled on the farmland outside the Kosovo complex; as for the tar, Radian notes without comment, that "this stream would be consumed in an on-site steam/power boiler or incinerator in the U.S. [to destroy the toxic compounds]. At Kosovo, this stream is landfilled."[99]

On the basis of this work, Radian concluded that, "*all* process units studied have a significant potential for polluting the environment."[100] The Radian study also found that the plant has adverse impacts on surrounding air quality. Analysis of about 3,000 samples[101] showed that the plant had an "unmistakable ... adverse impact"[102] on the air downwind of it, primarily because of "significant pollutants" from coal dust aerosols. "Ambient aerosol levels exceed the primary and secondary U.S. National Ambient Air Quality Standards," even though—as some have pointed out—the Kosovo plant "is located in a relatively remote area."[103] And the Kosovo air is not just dirty: the aerosols carry PAHs, probably released by the flare system, in very high levels. BaP exceeded EPA levels of concern (DMEGs) by a factor of 1,000, and "benzene probably exceeds the DMEGs by a factor of 10 to 100."[104] Thus EPA was led to conclude "that the measured concentrations of certain organic species [e.g., BaP] emitted from the Kosovo coal gasification plant may be sufficient to cause harmful health effects."[105] And, EPA's Ronald Patterson has warned, all this happened in a plant "only one-tenth the size of proposed U.S. facilities."[106]

Experiences in a German Lurgi facility and a South African Koppers-Totzek plant suggest that, even when properly controlled and subject to a considerable degree of governmental regulation, coal gasification plants remain potentially dangerous. The Ruhrgas AG plant in Dorsten, West Germany, was built in 1955 and now operates to process natural gas. Housekeeping and safety concerns are high, "in part a response to terrorist activities"[107] and also because "Ruhrgas operations are highly controlled by a number of government regulations and bureaus."[108] NIOSH visitors report that the plant is very clean, but accidents have happened, including an explosion in the iron oxide tray tower stemming from a leak in the oxygen injection line.[109]

Studies by the Tennessee Valley Authority, though, have suggested that Lurgi may be a more dangerous technology than NIOSH found. Because Lurgi wastewater contains unwanted byproduct tars, oils, phenols, NH_3, and naphtha, "along with a number of organic constituents," in significantly higher amounts than other gasifiers, it would "require more complex wastewater treatment."[110] TVA concluded that experience at operating Lurgi plants teaches that

> the large number of complex organic compounds present in waste and by-products of the Lurgi systems have a greater potential to adversely impact plant workers and the environment than wastes from other gasifiers.[111]

The AECI Ltd. plant at Modderfontein, South Africa, a Koppers-Totzek coal gasification plant that produces hydrogen for ammonia, provides another case study. NIOSH documents cite "a number of problems when the unit started up," including coal leakage, excessive erosion, "acute exposures to CO," and other gas leaks.[112] And worse:

> Modderfontein management is particularly sensitive to safety. During plant start-up there was a fatal accident. The design for the electrostatic precipitators for gas purification included improperly placed vents. A temperature inversion trapped gas at the top of the precipitator. When an analyst went to the top of the precipitator to sample the vents, he was killed by CO asphyxiation.[113]

In spite of the management's allegedly high degree of sensitivity to safety issues, the plant's workers suffer hepatomas, a liver tumor.[114]

TRW, Inc., has collected some meager data on the Modderfontein process and waste streams. While TRW announces that "the data from Modderfontein indicate that the streams tested do not appear to be of particular concern,"[115] only eight of the plant's more than 50 streams were properly tested. Still, testing showed that concentrations of phenols and cresols "consistently exceeded their DMEG [Discharge Multimedia Environmental Goals—EPA goals] values," along with PAHs in the Rectisol unit samples, and were thus "of concern."[116] A large number of metals, including iron, zinc, and lead, were also present at levels of concern.[117] However, "the only true discharge stream," the settling pond effluent (run-off), was much cleaner than the potentially dangerous process streams sampled.[118] This led TRW to conclude that "the wastewater treatment, consisting of a clarifier and a settling pond, was adequate to produce a final discharge that had lower pollutant levels than the fresh input waters supplied to the plant."[119] (It should be noted, however, that the "input waters" consist of recycled process water and heated sewage from Johannesburg.) No sampling or testing of the settling pond sludge was performed, although (given the compounds present, and their concentrations in the intermediate streams) one would expect it to be dangerous.

Both Krupp-Koppers (European K-T manufacturers) and the Tennessee Valley Authority have conducted further tests, not only at Modderfontein but also on two other K-T coal gasification plants, one in Kutahya, Turkey and the other in Ptolemais, Greece. On the whole, despite the Modderfontein problems, the K-T process seems to be a very clean one, primarily because of its high gasification temperatures. As Krupp-Koppers has suggested:

> A big advantage of the K-T process with regard to its effect on the environment lies in the fact that the produced raw gas contains no coal distillation products because of their spontaneous gasification at the extremely high temperature. Aromatics, phenols, and mercaptans have never been detected in K-T raw gas.[120]

Even K-T untreated wastewater "contains no tars, oils, or phenols,"[121] although it is high in NH_3 and suspended solids. The solids come from fly ash, which does appear to contain dangerous substances. It emerges from the existing plants in two forms: most is trapped in the raw product gases, where it is washed into a slurry and dumped in a settling pond; about 25-35% leaves as a granulated slag, which Krupp-Koppers says "can be deposited or used as a road construction material."[122] The settling pond effluent is either recycled, runs off, or is otherwise utilized: "in one commercial K-T coal gasification plant this waste water has for years been used to water the fields and as drinking water for animals."[123] (No reports on the fields, or the animals, are available.) As for the ash, what is not used for roads comes out as sludge, or as landfill, "both of which," according to the TVA, "present significant disposal problems."[124] In fact, "preliminary results of TVA's overseas test at . . . [the Ptolemais] coal gasification plant indicate that wet ash collection would remove certain gases from the product gas stream which may cause the wet ash to be classified as hazardous."[125]

CONCLUSION

The limited experience to date with commercial-scale coal conversion plants using modern control technology makes it impossible to be definitive in describing their likely environmental and health impacts. If the control equipment works as specified, then the impacts will undoubtedly be trivial. There is some reason to be skeptical, however, that the control equipment will always work as effectively as predicted, or even that it will function 100% of the time. In the event that the control equipment does not work properly, this chapter shows that there is some reason for concern.

We believe that the disposal of solid wastes poses the largest environmental threat from coal conversion, due to the unprecedented volume of wastes and to the fact that the solid wastes will contain the greatest quantity and concentration of the toxic constituents remaining after coal conversion. If these wastes reach water supplies through leaching or runoff, the ecological and public health consequences could be grave.

Water pollution from direct discharges by a coal conversion plant are expected to be less of a hazard, but this assumes that the strategy of zero discharge is successful, an assumption that may not always be viable.

The water requirements of a coal conversion plant, while large, are not likely in most cases to preclude the siting of a single facility. However, the water needs of a full-scale industry could create some serious conflicts with the needs of other users, and could force the price of water so high that some existing uses would no longer be economic.

The emissions of air pollutants from coal liquefaction and gasification plants are likely to be less than those of a conventional coal-fired power plant, and this air pollution is not a critical environmental issue. There is some concern, though, over the extent to which conversion plants will emit trace elements and other unregulated substances.

Assuming that the control technologies for the solid wastes, water and air emissions all work, the health effects of a coal conversion plant on the surrounding population are unlikely to be noticeable. No one is willing to project the health consequences under the scenario in which the controls undergo a massive or chronic, undetected breakdown. Living near a coal conversion plant may be comparable to living near a refinery or chemical plant, and thus may result in an increased incidence of cancer. Plant workers face a greater health risk than does the public at large, although again there is little conclusive evidence to support this piece of common sense.

The most important lesson to be learned from this chapter is that good monitoring systems are imperative: to watch for violations of the zero discharge system, to measure concentrations of pollutants in direct water and air emissions, to map concentrations of trace metals in water and soil, and to compile data on worker and local population health. An essential part of the monitoring effort is the collection of baseline data before the project begins operation. With a good monitoring system, it should be possible to make the plants run as cleanly as is typically promised in environmental impact statements.

5

Environmental Impacts of Oil Shale Development

The environmental concerns associated with oil shale development are similar to those of coal conversion. Oil shale contains many of the same constituents, including polyaromatic hydrocarbons (PAHs) and trace elements, whose release to the environment could cause serious human health and ecological problems. While oil shale retorting is less complicated than most coal liquefaction or gasification processes, far greater volumes of shale are required to produce the same amount of oil. In our judgment, the greater volume involved in shale production means that the task of preventing pollution is much more difficult than in the case of coal conversion. The impacts of oil shale production are also likely to be more concentrated geographically than those of coal conversion, due to the less dispersed locations of oil shale deposits.

The most serious environmental threat posed by oil shale development is created by the spent shale, which will be generated in volumes far greater than the solid waste produced by coal conversion. Spent shale disposal will entail, at a minimum, occupation of large areas of land, and may cause serious contamination of nearby groundwater and surface water supplies. As with coal conversion, the next most serious concern relates to water pollution from infiltration of evaporative and lagoon concentrates into the water system, or from backflooding of burned out *in situ* retorts. Air pollution probably poses the least significant threat.

SPENT SHALE DISPOSAL

The retorting of oil shale causes it to expand, with the result that the volume of spent shale will be 20 to 40% greater than that of the raw shale.[1] Even under conditions of maximum compaction, the spent shale will be 12% greater in volume than the original mined shale.[2] A 50,000-barrels-per-day oil shale plant will produce about 110,000 tons per day of spent shale, shale dust and coke, or about 10 times as much solid waste as would be produced by a coal liquefaction plant of the same production capacity. This quantity represents a volume of almost 2 million cubic feet per day, or enough solid waste to cover an area one mile square with two feet of spent shale each month.[3] Assuming a disposal pile height of 150 feet, the annual land requirement for such a plant would be 70-80 acres.[4]

These figures apply only to shale development using a surface retorting process. With *in situ* retorts, much of the spent shale can be returned underground. However, *in situ* retorts produce a greater volume of spent shale than do surface retorts because the recovery rate is lower and leaner shales would be used in the retort.[5]

The spent shale problem is aptly summarized in the following statement from the Department of Energy's study, *Synthetic Fuels and the Environment:*

> [Retorted shale] represents a major solid waste management and disposal problem from the surface and modified in situ operations from both the amount and its content.[6]

At best such volumes of solid waste will create a tremendous visual impact, which one environmental impact statement describes as follows:

> The most significant modification of land surface will be the processed shale disposal embankment which will cover 800 acres of Davis Gulch to a maximum depth of 600 feet, with an average depth of approximately 350 feet. The significant topographic alteration of this area will create an undulating plateau-like area, in contrast with the present ridge and valley regime. The impact of the processed shale embankment will be substantial in terms of magnitude, permanence, and importance.[7]

In addition to the massive topographic disruption that will be caused by spent shale disposal, there are serious concerns about the stability of the waste piles. Improper pile design may cause erosion and/or landslides. More significantly, toxic substances in the spent shale could be carried into the water supply by runoff or leachate.

It is hoped that revegetation of the piles will stabilize them structurally. However, the extent to which spent shale can successfully be revegetated is somewhat in question. With surface mining of coal, reclamation and revegetation are typically attempted by restoring the overburden in as close to its original condition as possible. Spent shale does not by the wildest imagination approximate soil. Perhaps the most likely route for revegetation of spent shale, therefore, will be to harden the shale by using some sort of chemical spray or by adding such materials as asphalt or limestone, and then to cover it with a "friable soilcover." In somewhat of an understatement, the Office of Technology Assessment concluded that the use of a soil cover is preferred, because

> . . . the chemical and physical properties of processed shale make it much less amenable to supporting plant growth that resembles the diversity and density of the present natural vegetation ecosystems.

The following synopsis of spent shale characteristics suggests why this might be the case:

—Shales that have been retorted at high temperatures will, when moistened, harden "in a reaction similar to that which takes place in cement;" "If shale hardened by this process were to be exposed by erosion, it might prove impenetrable to moisture and plant roots;"

—"If processed shales are to be used directly as a growth medium, their alkalinity must be reduced;"

—"Spent shales have been shown to be highly deficient in the forms of nitrogen and phosphorous available to plants;"
—"High concentrations of salt in the soil media [spent shale] restrict water and nutrient uptake;"
—"[A]bout 5 acre-feet of water per acre will be needed for leaching [to remove salts from shale] and plant growth;"
—The temperature at which the spent shale will enter the disposal site, over 98 degrees Fahrenheit, will create a heat reservoir.[9]

Field studies have been attempted on revegetation of spent shale from TOSCO II, Union and Paraho retort wastes. It is reported that,

> These studies show that *with intensive treatments* plant growth can be established directly on spent shales although use of a soil cover is more successful."[10] (Emphasis added.)

However, another report warns that this research "has been done on a small scale and has not addressed the question of long-term revegetation and stability."[11] Until evidence is obtained that spent shale can be revegetated under natural conditions, developers should be pressed for a commitment to continue maintenance of the revegetated area for as long as may be necessary (which may be decades).

Even if the spent shale is successfully revegetated, the toxicity of the wastes gives rise to several other concerns. The solid wastes from oil shale retorting include raw mined shale, overburden where stripmining occurs, raw and spent shale fines from dust control devices, and other process wastes such as spent catalysts, water treatment sludges, oil sludges and shale coke. All of these wastes contain substances whose introduction to the environment could be harmful to people and wildlife. Spent shale represents by far the largest solid waste stream. In addition to salts and organic substances, it can be expected to contain the trace elements mercury, lead, cadmium, arsenic, copper, zinc, selenium, iron and boron. While these substances are toxic in small quantities to humans and animals, they are not all toxic to plants. They may thus be taken up by the plants and then ingested by grazing animals. It may,therefore,be necessary to restrict wildlife, including small burrowing rodents, from the revegetated area.

Probably the most serious concern regarding spent shale disposal is the potential for leaching of the toxic elements in the shale into the water supply.

While revegetation may control erosion from the top of the piles of spent shale, rainfall or irrigation water percolating through the piles may carry significant quantities of heavy metals, such as cadmium, lead and arsenic, into the groundwater. In order to prevent such movement of toxic substances, it is proposed that at least the initial groundcovering layer of spent shale will be formed into an impermeable liner, by wetting it down and compacting it. Some tests indicate that this approach can be effective. For example, one test indicated that percolation of water through a 4.5-foot thick test bed of compacted shale required the equivalent of seven inches of rain per day for a week.[12]

Other assessments of the ability to make spent shale impermeable are less optimistic. EPA-sponsored field studies found that the permeability of compacted, direct-mode Paraho spent shale, for example, is quite high.[13] EPA concludes that,

> All that can be stated with certainty at present is that any proposed spent shale disposal providing for compaction to make the shale impervious should first prove that the particular spent shale and compactive effort can indeed make the spent shale impervious.[14]

The EPA report goes on to state that,

> "Even though an impervious layer (30 cm/year or 12 inches/year or less permeability) is installed below a disposal pile, there is a risk of groundwater pollution if the groundwater flow rate below the pile is too low to provide substantial dilution of any leachate that eventually seeps through the impervious bottom of the disposal site."[15]

Questions regarding the long-term stability and permeability of the spent shale led DOE to conclude that,

> "Stabilization and management of spent shale and other wastes from surface retorting processes will require additional RD&D before the leaching concerns are resolved."[16]

Sizeable quantities of leachate will be produced from the spent shale. Sources of leachate include precipitation, water used to cool the shale, water needed to compact the shale, water used to leach the shale before revegetation, and irrigation water once revegetation is being attempted. Spent shale leachate can be expected to contain all the trace elements, including boron, lead, nickel, selenium and strontium. It will also contain salts and organic compounds, although the latter are "not well-defined at present."[17]

WATER POLLUTION

In addition to the potential pollution from leaching of spent shale, oil shale conversion can result in water pollution due to deliberate or accidental discharge of wastewater streams created at several different points in the process. These include mining, retorting, upgrading, air pollution control, steam generation and water cooling.

The main impact on water quality from the introduction of spent shale leachate would be an increase in the salinity of the receiving water. EPA described the problem as follows:

> ". . . Leachates from spent shale constitute a potential source of salt loading to surface water owing to the proposed disposal methods and the physical and chemical characteristics of spent shale."[18]

Water is used in mining to clean the air of dust and gases. Dewatering of shale deposits also produces a dirty water stream that must be cleaned and/or disposed of. The retorting process itself results in the release of water vapor from the shale, which forms a dirty condensate. *In situ* retorts will generate substantially larger quantities of polluted water than will surface retorts. The scrubbers and water sprays used to reduce air pollution produce liquid streams that are loaded with chemicals, particularly sulfur and ammonia recovery units. Water used in the boiler to produce steam becomes built up with dissolved salts and eventually must be disposed of, as is also the case with recycled cooling water.

For a 100,000 barrel per day facility, for example, the stream of wastewater will be about 3,500 gallons per minute, a volume greater than the amount of oil to be produced. Given the large volume of wastewater resulting from oil shale production, its contents are of some concern.

The retort water is the most polluted of the waste water streams. The levels of ammonia and boron in untreated retort water are sufficiently high to be toxic to fish and other aquatic life.

The wastewater streams are too high in concentration of salts and toxic substances to return to a natural body of water without treatment. In some cases, it is envisioned that the wastewater will be disposed of entirely in the spent shale pile, where it can fulfill the function of compacting the spent shale to make it impermeable, as described above. This approach, of course, increases the toxicity of the spent shale, which may become a problem if the compaction method proves not to render the spent shale totally impermeable. One study found that using retort process water to moisten spent shale would increase its content of organic matter by over 70%. It is, therefore,anticipated that such a disposal method "will appreciably increase the potential release of organic wastes, including carcinogens."[20]

Three other alternatives are available for wastewater disposal. The first is to treat it, using one of a number of chemical and and physical processes, producing a small volume of solid waste by-products, with some undesirable chemical properties, which will be disposed of along with the spent shale. While these treatment processes have been proven in similar industries, such as oil refining, they have not been tested extensively on oil shale operations. In particular, it is not known how effectively these methods will function when more than one must be applied sequentially to one or more wastewater streams. It is expected, however, that these processes will remove the majority of the ammonia, hydrogen sulfide and carbon dioxide present in the wastewaters.[21]

It is unlikely that a developer would propose to discharge even treated wastewater to a river or undergound aquifer. Most of it will be used for dust control or spent shale compaction, as mentioned above. The disadvantage of this approach was pointed out by EPA:

> In essence, this converts a point source of pollution, which would be highly regulated under existing laws, to a non-point discharge, which is not well-regulated at present.[22]

The second approach for disposal of wastewater is via evaporation, which can be accomplished by ponding the waste water, allowing the salts and other chemicals to accumulate in a shallow basin. This approach requires that care be taken to prevent rainfall or flood runoff from carrying the toxic substances from the pond to nearby water bodies.

The third option for disposal of wastewater is to reinject it into a mine after treatment. Reinjection is legal only if the quality of the discharge water is at least as good as that of the underground water likely to be impinged by the discharge. Mine drainage water will probably be disposed of by reinjection, at least until commercial-scale production is begun. It is assumed that such a practice will have no adverse impact on the groundwater regime, although,

> . . . water quality could be degraded because of the increased ground water flow, the exposure of new mineral surfaces by fracturing, and the changes in undergound microbial populations. [23]

In sum, the *expected* impacts of oil shale conversion on water quality are minimal. It is anticipated that some increase in the Colorado River's salinity may occur as a result of oil shale development, but that most of the toxic elements will be removed from the wastewaters and disposed of with the spent shale. In order to assure that such predictions are borne out, however, local and regional water quality should be continually monitored for breaches caused by leaching of the spent shale piles and/or discharges of untreated wastewater. Since surface and groundwaters intermingle, contamination of one supply can spread to the other. Thus, monitoring of both surface and groundwater is needed. A good discussion of useful elements in a monitoring program is found in the OTA report cited earlier.[24]

WATER REQUIREMENTS AND AVAILABILITY

Oil shale retorting requires water for moisturizing the raw shale, product cleansing, cooling, and to stabilize spent shale piles. Water use for spent shale stabilization is by far the largest of these requirements, with half or more of the water used by an oil shale facility going to this purpose. Oil shale facilities will tend to use more water than coal liquefaction facilities of the same size.

The discussion in Chapter 4 of water constraints on coal conversion is also applicable to oil shale development. Most studies agree that, on an average flow basis, water supplies are adequate for a sizeable oil shale industry. For example, the OTA report on oil shale found that surplus water legally available to Colorado, Utah and Wyoming is sufficient to support a 2.1 million barrel per day industry through the year 2000.[25] However, the analysis concluded that some new reservoirs would be needed.

The OTA report assumes that oil shale development will occur primarily on the Green River, White River and Colorado River subbasins of the Upper Colorado River basin. While it estimates that a 1 million barrel per day industry would increase total water consumption in the three subbasins by only 10%, White River use would increase by 150%.[26] New reservoirs would be needed on the White River to meet the industry's demand, and river flows would be "substantially reduced."[27] The main stem of the Colorado River may also need new reservoirs in order to supply the growing needs of both shale development and agricultural irrigation.[28]

As we pointed out in Chapter 4, reservoir construction is not without its environmental impacts, including losses of water due to increased evaporation. These losses should be charged to the oil shale development. It should also be kept in mind that river flow estimates are not always applicable to specific sites. If actual site flow is less than the river average, aquatic impacts may be greater than those predicted by the developer.

AIR POLLUTION

Oil shale projects potentially can emit large amounts of air pollution. Most of the pollutants released from oil shale mining and retorting are of the "conventional" sort, which are generated from combustion of any fossil fuel and can largely be controlled by available technologies. There is also some concern, however, that significant levels of trace elements may be emitted to the air during oil shale conversion.

As discussed in the preceding chapter, federal air quality standards apply only to the "conventional" pollutants. If they comply with the standards for these pollutants, individual shale conversion plants should be relatively free of visible pollution. But this generalization may not be true of the large, concentrated shale industry proposed for Utah and Colorado:

> The development of a large oil shale industry . . . will degrade the air's visibility and quality. Even if the best available control technologies are used and compliance is maintained with the provisions of the Clean Air Act, its amendments, and the applicable State laws, degradation will occur.[29]

The "conventional" pollutants generated by oil shale operations are particulates, SO_2, NO_x, hydrocarbons, and carbon monoxide (CO). These pollutants originate in mining and preparing the shale, retorting, upgrading and stripping the shale of ammonia and sulfur, auxiliary processes, and treatment and disposal of spent shale. Mining can release hazardous substances such as silica, salts, mercury and lead during blasting. Large quantities of dust may also be released during mining and during storage, transport and crushing of the shale. The combustion of fossil fuels to heat the retort will produce SO_2, NO_x, CO and hydrocarbons, and the gases produced in the retort will contain hydrogen sulfide (H_2S), carbonyl sulfide (COS) and carbon disulfide (CS_2). Some polycyclic organic matter will also be produced from pyrolysis of the shale, as may certain trace elements. Upgrading the shale and cleaning of associated gas products will generate carbon disulfide, carbonyl sulfide, SO_2, H_2S, ammonia and hydrocarbons.

A modified *in situ* (MIS) retort will generate substantially smaller volumes of pollutants, since the undergound operations offer a greater inherent control of emissions. In surface retorts, the greatest sources of particulates are in shale preparation and the actual retorting of the shale. With modified *in situ* retorts, by far the greatest source of particulates is in mining the shale. The bulk of the sulfur dioxide will stem from the ammonia and sulfur recovery units, with the surface retort producing about a third more than the modified *in situ* process. Nitrogen oxide generation will be higher in the MIS process than at the surface retort, largely due to emissions from steam and power generation. In surface retorts, hydrocarbons will come mainly from the retort, followed to a smaller degree by product storage and mining, both of which are responsible for generation of significant amounts of hydrocarbons under the MIS process. Carbon monoxide will be generated primarily in the mining stage for both surface and MIS retorts, although the surface retort will also be a significant contributor.

In practice, emissions will be controlled to a fraction of the pollutants generated. Water sprays, cyclones and scrubbers will be used to capture dust. Baghouse filters and electrostatic precipators can control the emission of the microscopic particulates. Chemical processes are used to remove sulfur and nitrogen from the product gases.

HEALTH IMPACTS

As in the case of coal conversion, the lack of experience with a commercial-size oil shale facility using "best available control technology" makes it difficult to state conclusively whether any health impacts to workers or the surrounding population will occur. But as is also true for coal conversion, mining and processing, the shale will release a number of compounds and trace elements known to be harmful to man. The extent to which oil shale plants adversely affect human health will depend largely on how well the various pollutants are controlled. Thus, active citizens participation in establishing the permit conditions of a new facility is desirable. Of similar importance is the establishment of a program to monitor its acute and long-term health effects.

Mining of oil shale is likely to have similar health and safety hazards to those of coal and other mining. The danger of mining accidents could be somewhat increased over that in other industries due to the fact that the oil shale mining rooms will be larger, requiring more internal support. While in practice the accident rate has been lower for oil shale than for other types of mining, the experience thus far has been limited to experimental mines. Mining at modified *in situ* operations may pose unique hazards to miners due to high temperatures and fires.

The more serious problem associated with mining is worker inhalation of dusts. Exposure will also occur during handling of spent shale. The dusts from oil shale contain silica dusts, which "have been the single greatest health hazard throughout the history of underground mining."[30] Associated with inhalation of oil shale dust are silicosis, "shalosis," and bronchitis. ("Shalosis" is a lung disease related specifically to oil shale dust exposure.)

The OTA report summarizes the results of several studies on effects of dust inhalation. One study found chronic bronchitis among oil shale miners to be 2 to 2.5 times more prevalent than in the general population. Other studies were less conclusive. For example, the National Institute for Occupational Safety and Health (NIOSH) examined the death certificates of 167 oil shale workers and failed to find any relationship with respiratory diseases. However, OTA pointed out that the small numbers of workers on which these studies were based, coupled with the limited exposure of the workers involved, makes their results less than conclusive.[31]

Mining and handling oil shale may also expose workers to carcinogenic PAHs and trace elements. The NIOSH study mentioned above found that the percentage of oil shale workers who died from colon and respiratory cancers exceeds that in the population of Colorado and Utah males,[32] possibly as a result of the workers' exposure to carcinogenic substances.

Crude shale oil contains many hazardous substances, such as chlorine, sulfur, nitrogen, and heavy metals, including a high level of arsenic. Refining generates more toxic contaminants: hydrogen sulfide, hydrogen chloride, hydrochloric acid, sulfur dioxide, sulfuric acid, nitric acid, nitrogen oxides, mercaptans, carbon monoxide, and benzene. PAHs from retorting and refining oil shale are "a major potential health hazard" for workers.[33]

As discussed in Chapter 4, PAHs have been shown to induce cancer in animals. Studies of effects on humans have been limited in scope. In the early 20th century, Scottish oil shale workers were reported to have a higher risk of developing skin cancer than people not exposed to oil shale.[34] At the same time, shale-derived lubricating oils used in textile factories began to be linked to various types of cancer.[35] The incidence of skin cancer among females employed in the Russian (Estonian) shale oil industry was significantly higher than in the general population.[36] Though the studies of the Estonian oil shale industry have shown an association between oil shale processing and cancer in both the workers and local residents, they contain little information on the working conditions or ambient concentrations of shale-derived pollutants in the vicinity of the plant, making it difficult to relate these results to the U.S. oil shale industry.

An additional occupational risk to oil shale workers is created by the potential synergistic relationship between sunlight and substances in oil shale, which may increase the chance of skin cancer (especially on the Colorado Plateau where the sunlight contains higher levels of ultraviolet radiation than at lower elevations).[37] This type of synergistic effect between ultraviolet radiation and volatiles from coal-tar has been discovered to cause skin cancer.

The health risk to the population surrounding an oil shale facility is even more elusive than the risk to workers. As discussed in Chapter 4, there is some evidence linking residence near heavy industrial and chemical facilities to increased cancer rates, and there is some similarity between the emissions of such facilities and those of an oil shale operation. One study of the local population near the Estonian oil shale industry, which has been active for the last 40 years, found an "unusually high" incidence of stomach and lung cancer. We do not know, however, enough about the conditions of those operations to extrapolate the results to U.S. oil shale development.

Given the potential for workers and local residents to be exposed to the known toxic and carcinogenic components of oil shale, the best health insurance policy will be the establishment of a good monitoring system. As some of the hydrocarbons produced by oil shale development may be more biologically active than those found in conventional petroleum processes, new sampling equipment may need to be developed.[38] A good discussion of a comprehensive approach to evaluating the health effects of oil shale development is found in an inter-agency report entitled, *Health and Environmental Effects of Oil Shale Technology.*[39]

CONCLUSION

The most pressing environmental problem resulting from oil shale development is the disposal of spent shale, which will require the disturbance of large land areas and has the potential to pollute ground and surface waters. Direct water discharges and air emissions are of lesser concern, although monitoring is needed to ensure that federal standards are indeed being met. Careful monitoring of workers' and nearby residents' health should also be maintained.

6

Socioeconomic Impacts

Some communities might be willing to tolerate the environmental damage caused by synfuel development in exchange for the economic benefits a project may bring. However, many energy developments have proven to be a mixed blessing even in this regard. The economic consequences of industrial development reach far beyond the obvious increases in employment. New schools, hospitals, and sewers are suddenly required. Living costs rise. A police force becomes necessary. Lifestyles change.

Most of the socioeconomic consequences of synfuel development stem, either directly or indirectly, from the massive increases in population that occur in the vicinity of a new plant while it is being built. The impact is greater in rural than in urban areas because population increases are proportionately larger. Many areas targeted for synfuel development, especially in the West, are just such rural areas. A Department of Energy study concluded that 22% of the 324 counties with sufficient coal to support synfuels development had populations too small to assimilate such development without severe dislocations.[1] In the sparsely populated West, the percentage is higher. In the Northern Great Plains Region, for instance, only two counties had populations exceeding 20,000 in 1970.[2]

Generally, local residents will have the appropriate skills for only about 20% of the jobs created at a new synfuels plant, mostly those in the less skilled and lower paying categories.[3] In the first year of construction of a gasification plant, the work force is composed almost entirely of professional and technical workers, mainly engineers.[4] According to one government study, during construction of most energy conversion (synfuel) facilities, "few workers without the relevant experience or training could be hired."[5] Apart from engineers, construction requires hundreds of pipefitters, electricians, insulation workers, mill workers, and ironworkers. This demand for skilled workers who are unavailable locally leads to a large influx of workers.

The size of the incoming labor force is determined by the type of synthetic fuel being developed. The peak work force required for construction of a 45,000 barrel/day indirect liquefaction facility is 10,000 workers.[6] Construction of a high-Btu gasification project of equivalent capacity requires about 4,000 workers.[7] However, the magnitude of the population increase is not adequately reflected in these figures. Jobs bring job hunters. Married employees bring their families. Families create a demand for schools, hospitals, stores, etc. In fact, each person *directly* employed by a synfuels plant brings from 4.5 to 7 other people into the area around the project.[8] The total population inflow for an indirect liquefaction plant was estimated at 46,000 people.[9] In other words, small tons and counties will be dwarfed by synfuel employment.

The problems presented to communities by the massive influx of workers during the construction phase of a project are inverted during operation, because operation of the plant provides considerably fewer jobs than construction. Labor requirements may be as low as 10% of the construction work force, and generally will be less than half of the peak construction force. One 1980 employment projection for indirect liquefaction plants placed the third year construction work force at 10,000, while the fifth year work force is expected to be only 800.[12]

This steep decline in jobs once construction is finished contributes to part of the traditional "boom-and-bust" cycle which plagues energy-impacted communities. Municipalities must face a sudden oversupply of housing and community services that they can no longer finance. The "bust" is discussed in greater detail later in this section.

The phenomenon of a sudden flood of skilled workers, followed in a few years by a sudden multiplication of the unemployment rate, places severe stress on public services, municipalities, and social institutions. Change comes far more quickly than small towns can respond to it. Past experience has demonstrated all too clearly the problems that must be faced. The rest of this section deals with the potential social and economic problems that may beset a community as a result of the massive influxes of synfuel plant employees, their families, and support systems. We will also discuss methods for assessing the potential effects on a specific community and ways to reduce the negative effects on a community.

BURDEN ON COMMUNITY SERVICES

The rapid expansion in population associated with a new synfuel facility creates a sudden demand for community services. The new population needs housing, medical services, and schools, all of which require time and money to develop.

The demands placed on a community are immediate. Roads must be built, *now*, houses provided for workers, *now*, classrooms for new children, *now*. But the revenues are not immediately forthcoming. The plant often cannot be incorporated into the community tax base until it begins operation.

Many of these services could be delivered if given adequate funding, but in most cases the time lag between the period when the greatest funds are needed and when the plant begins operation, and therefore tax payments, is three to four years. It is this time lag that causes many of the problems in boomtowns. To lessen adverse impacts, communities need funds two to three years prior to the intensive plant construction period, when massive population influxes began.[13] According to one government report, the "school classrooms, utilities, roads, etc., may need to be constructed, paid for and abandoned before the new plant is ever entered on the local tax rolls."[14]

The Rocky Mountain Oil and Gas Association concluded that $150 million in capital improvements for public facilities will be needed during the next 25 years in Rio Blanco County in the Piceance Basin alone to accompany the development of four large oil shale facilities. Less than $3 million was available in 1981, however, when the needs for the region were estimated at $7.5 million.[15] A second study conducted in Colorado concluded that although three counties in the oil shale region of the state would eventually show a positive financial return, it would take seven years before the operating surplus exceeded the initial deficits.[16] Thus, in the early years when the socioeconomic costs are the highest, small rural communities will be subsidizing the booming synfuel industry.

The front-end financing problem is further complicated when new tax revenues generated by the synfuel projects go to state or county governments and not directly to the communities where the in-migrants settle. Although synfuel development causes an increase in assessed valuation and an accompanying rise in property tax revenue, the taxing jurisdiction of the synfuel site may be distinct from population centers. One federal government analysis concluded that these jurisdictional mismatches are likely to be a frequent occurrence.[17] The report observed that:

> energy developments will very seldom be within the corporate limits of cities and towns. Frequently they will be located just across school district, county, or even state boundaries from where the new population will live. . . . Where population growth occurs in one taxing jurisdiction and the increases in assessed valuation are received by another jurisdiction . . . the population-impacted jurisdiction is almost certain to have serious financial difficulties.[18]

The effects of the funding gap will be felt in all areas of public services: housing, health care, schools, fire and police protection, and others.

Housing

Many energy boomtowns have experienced immediate and dramatic increases in the cost of housing, lots, and rents. Within one year, for example, the price of a lot in Valdez, Alaska, shot up from $450 to

$10,000 due to a massive influx of construction workers for an energy project.[19]

Unfortunately, it is extremely difficult for communities to alleviate the housing shortage. In most communities mobile homes take up the slack, often providing as much as 30% of the housing in energy boomtowns.[20] These mobile home "developments" often cluster about the fringes of the community—"aluminum ghettos" plagued by lack of roads, sanitation systems, power, and water.

In the boom community of Craig, Colorado, even mobile home settlements could not accommodate the rapid growth. An uncounted but significant number of in-migrants, possibly several hunded, lived in RV trailers in the campgrounds outside of Craig. Lack of essential facilities threatened to cause health problems.[21] Recently, at a synfuels site in Garfield County, Colorado, construction workers were found camping on river banks and highway rights-of-way without adequate water, sewer, or sanitation services. County commissioners were so alarmed that they threatened to revoke the construction permit granted to the synfuel sponsor, Union Oil of California.[22]

Health Care

The growing population may place a great strain on medical services, particularly in rural areas if the concentration of doctors is already low. From 1970 to 1974 Rock Springs, Wyoming, underwent rapid growth due to the construction of a coal-fired power plant and the expansion of mining operations in the region. The county population more than doubled in four years, an annual growth rate of 19%.[23] In 1970, there was one doctor for 1,800 people. Four years later, one doctor served 3,700. This figure was almost three and a half times worse than the state average ratio of one to 1,100 and more than six times the nationwide average of one to 612. Consequently, health care was a major problem; 40% of Rock Springs residents were forced to seek care elsewhere.[24]

Education

A rapid rise in school age population is likely to accompany the increase in worker population resulting from synfuel development. It is estimated that there will be one new student for every new energy worker.[25] School systems may require dramatic and costly expansion in order to handle the sudden increases in school age population.

Because energy projects ordinarily do not provide tax revenues until production begins, a school district must drastically hike residential taxes to try to accommodate the new demands in the meantime. In Wheatland; Wyoming, a town which experienced a boom from power plant construction, the school district increased its taxes five-fold.[26] Even after the industry begins to pay taxes to the county and state during operation, school districts may still be deprived of synfuel revenue if the plant is located outside the affected district.

Boomtown schools can also experience social problems that no amount of money can solve. Transiency is particularly disruptive both socially and acaemically for the educational system. For example, when Wheatland finally succeeded in hiring an ample number of teachers and enlarging its facilities, it then saw 454 of its 934 elementary students enter or leave school in midyear.[27] The sudden influx of new students from different socioeconomic backgrounds, holding different values and following different lifestyles, often creates social and psychological conflicts. Studies have shown that with these new social pressures come the increased incidence of drug use, truancy, and delinquency.[28]

Fire and Police Protection

The sudden change from a small, stable population to a booming growth center may be accompanied by a rise in crime. Existing fire and police personnel may be unable to cope with the needs of a rapidly growing community. Again, the foremost reason is the initial shortage of funds. Police budgets have been known to double during a boom period.[29]

Other Public Services

The impact of rapid development and population increases without accompanying increases in funds to pay for improvements can result in a deterioration of other important services. County roads (when they exist) are often inadequate to support thousands of workers commuting daily. The transportation of construction materials, heavy machinery, and coal adds to the overload. Other transportation facilities, such as railroads, which once served farmers and ranchers, may also be taken over by mining and industry. The increased use of existing railroads for coal hauling can severely disrupt normal and necessary community activity.[30] The increase may be dramatic. The train traffic between two small coal boomtowns in Colorado, Craig and Bond, has been projected to increase from four to 30 trains per day due to new federal coal leasing.[31]

The rapid increase in community housing and the lack of revenue for upgrading facilities can also lead to inadequate sewer, sanitation, and water service.[32] The maintenance of fire pumping capacity may be a serious problem in arid western towns.[33]

INDIRECT CONSEQUENCES

Some of the impacts on community services and housing discussed above can be mitigated if adequate funds are available and if communities carefully plan their growth. Other effects, however, are more difficult to deal with.

Inflation

Construction of a new synfuel facility may cause a local rate of inflation substantially higher than the national rate. One cause of boomtown inflation is the great disparity between the wages offered in the mining and construction sectors and those offered by existing local businesses. The Wyoming Employment Security Commission reported in 1978 that mining employees earned an average weekly salary of $471, while employees in agriculture, forestry, and fisheries (originally the economic staple of the region) earned only $168 weekly, and those in retail trade only $149 weekly.[34] This polarization of incomes may become even more severe, if a company is forced to increase wages to draw skilled laborers into an isolated area.

Ironically, inflated wage rates deprive local employers of workers, even when there is unemployment. And, as wages increase, so does the cost of living. Shortages in housing, goods, and services can drive prices even higher. Hardest hit will be those residents on low or fixed incomes—the poor, elderly, single women, and pensioners. The Colorado Department of Local Affairs has reported that the sharp rise in living costs caused by the influx of workers and the economic activity associated with oil shale has already severely affected senior citizens in northwest Colorado.[35] While some prosper from a boom, many others will suffer, their quality of life assailed by higher costs as well as the other boomtown afflictions.

Unemployment

Synfuel plants are often touted as a cure to local unemployment. But the efforts of a project sponsor to attract construction workers to its site may cause large numbers of hopeful employees to flock to remote communities. Statewide advertising recently attracted 1,800 applicants in one week for 200 job openings on a shale oil project, near the small town of Parachute, Colorado.[36] Energy construction projects have repeatedly *raised* local unemployment rates. For instance, the unemployment rate in Whatcom County, Washington, rose 6-9% during the construction of an oil refinery.[37] Similarly, the construction of the Alaska Pipeline boosted the unemployment rate in Fairbanks by over 11%, because it attracted far more job seekers than it employed.[38]

Ranching and Agriculture

The economic benefits resulting from synfuels development will be concentrated in and around towns where workers spend their money. Much of the disruption, however, will be felt by surrounding farmers and ranchers. Not only will mining impacts occur, but the land requirements for a plant and its ancillary facilities are enormous. Land is needed for waste disposal, roads, railroad spurs, and pipeline and utility corridors. New highways will cross farms and ranches, consuming 25-30 acres per mile.[39] Bisecting previously contiguous tracts, these highways can damage farming efficiency and may necessitate costly fencing for ranchers. These impacts will be centered in geographically isolated locations while the energy consumption market is hundreds of miles away. Many of these land use changes are permanent, outlasting the 20-25 year life of a synfuel plant.

Farming and ranching enterprises may be adversely affected by synfuel development due to a resulting shortage of labor. One recent government study bluntly noted that "the basic economics of coal development are adverse to agriculture." Agricultural operations, already at the margin of solvency in many cases, may not be able to compete directly with an energy corporation for labor. Farmers and ranchers may also be hurt by the increased cost and scarcity of support services. During a labor shortage caused by the construction of a coal-fired power plant in the town, a farmer in Wheatland, Wyoming, reported spending an entire day during planting season looking for a mechanic to replace his clutch. Unsuccessful in that endeavor, he finally had to rent a new pickup.[41] Because agricultural products often are exported from the local trade area for sale in competitive regional or national markets, farmers and ranchers may be unable to pass their higher costs on through the price of their products, and thus are very vulnerable to dramatic increases in labor costs.

Local Business

Local businesses may not benefit as greatly as might be expected from the inflow of population and cash from a synfuel project. Past energy boom developments have led to the introduction of chain businesses in rural areas. New supermarkets, fast-food restaurants, and other franchises located in newly constructed shopping malls may compete with or reducing the profit of existing businesses.[42] Local retailers and manufacturers may have trouble hiring and keeping employees as they find themselves unable to compete with the wages and salaries offered by mining and construction companies.

Newcomers may choose not to shop at local establishments. If a larger town is within reasonable driving distance, studies have shown that the majority of new residents will conduct a substantial portion of their shopping outside the community. This pattern can have a far-reaching, destructive impact. The loss of newcomers' disposable income deprives the boomtown of sorely needed sales tax revenues, exacerbating the front-end financing problems that the community may be experiencing. Projects and their employees therefore may contribute considerably less than would be expected to the local business economy and the local service sector.

SOCIAL CONSEQUENCES

The type of rapid, temporary development associated with a synfuel project located in a small, isolated community can produce social tensions as well as shortages of essential services. The social stress experienced in energy boomtowns have been well studied. Studies have shown that as the rate of growth increases, so does the social chaos experienced in boomtown communities.[43] Social scientists agree that most communities experience serious social strain when the population growth rate exceeds 4-5% per year.[44]

The Gillette Syndrome

The Gillette Syndrome was named for the social consequences of boomtown life as they were observed in Gillette, Wyoming, in the 1970s. As a result of energy development, Gillette grew from 4,000 people in 1969 to 14,000 in 1981.

The Gillette Syndrome describes the problems of boomtown life. One social scientist graphically described boomtown conditions in this way:

> Divorce, tensions on children, emotional damage and alcoholism were the result. Children went to school in double shifts; motels turned over linens in triple shifts. Jails became crowded and police departments experienced frequent changes in personnel in the tradition of frontier justice. Out of frustration with the quality of living, it appeared that mayors shuttled in and out of office like bobbins in a loom. Depression was rampant with suicide attempts at a rate of one per 250 people.[45]

Craig, Colorado, also the site of large-scale energy development, saw its population double in two years. Meanwhile,

—crime against property increased 220 percent;
—crimes against persons increased 900 percent;
—family disturbances rose 250 percent;
—child behavior problems went up 1000 percent;
—alcohol-related complaints increased 550 percent; and
—drug-related reports increased about 1400 percent.[46]

Sadly, Craig's experience was not unique. Other communities have fared worse.[47]

Women and Families

Boomtown life can be especially hard on wives of in-migrant laborers. Often isolated by their residence in mobile homes at the fringe of town, they are frustrated by a lack of recreational activities and essential public services for themselves and their children. Often, local health systems may not be prepared for the increased demand for obstetrics, gynecology, and pediatrics.

Divorce rates climb rapidly in boomtowns. In Rock Springs, Wyoming, the ratio of marriages to divorces was 1.8 to one during its boom while the ratio of a nearby undeveloped county was 3.3 to one.[48] Another boom area, Moffat County, tripled its number of divorces during boom years.[49]

Family deterioration is also revealed through the rise in child abuse. Again Moffat County is illustrative. The number of reported cases in 1978 rose from nine to 46, more than fivefold, and again far more than sheer population increase would warrant.[50] Spouse abuse has also increased in boomtowns.[51]

Long-Time Residents

Rapid growth leads to other lifestyle changes disturbing to long-time residents. The first is the loss of and helpfulness among people who need each other for mutual survival. Second, local residents may feel that they have lost control over their community: newcomers take over leadership positions; the federal government and multinational corporations control the town's economic destiny. The result can be a loss in community pride.

The elderly of small communities can be seriously hurt by rapid new development. They must confront the same boom-generated problems as all residents, but are usually far less able to handle these crises. The rapid changing of the pace, people, and physical layout of the community is likely to affect the elderly profoundly. The U.S. Department of Housing and Urban Development described their problem:

> Most of the elderly have been in the community for a long time, and counted on a quiet retirement in a community they knew. With the rapid growth, the community is no longer quiet and may not be recognizable.[52]

Political Structure

The advent of large numbers of in-migrants to rural towns can mark the demise of the existing political system. Long-time residents lose political control of their community to the new population and the leaders of industry. The tendency for incomers to fill religious and elected civic positions has been documented.[53] When one considers the size of the incoming labor force and the size of the original voting populace, it is easy to see how political power and philosophical dominance can slip from local residents. This shift in political power can aggravate tensions between the new and old residents.

THE "BUST"

Colorado Governor Richard Lamm has dubbed the energy-rich, arid west "the Match Belt." In contrast the the Sun Belt and the Frost Belt, western energy towns flicker to life, burn brightly, and then are snuffed out.[54] In the Match Belt, economic depression and rapid out-migration of residents are likely to follow the years of intense economic activity.

A town may experience a major bust once a synfuel project comes to the end of its lifetime, usually predicted to be 25 to 30 years. A less dramatic bust may occur earlier on, when workers leave after construction of a project is complete.

A significant drop in population may make a community unable to support the public facilities and services that were established to serve the larger population. Expensive facilities, once manageable in the "boom" period, might be burdensome to a town with a newly shrunken tax base. The bust might also generate a housing surplus, negatively affecting the housing, real estate, and local construction markets.

Appalachia has borne the brunt of many cycles of boom and bust. When coal production declined in the 1950s, low wages and unemployment prevailed, and the limited incomes meant limited services. Consequently, education, health care, sanitation facilities, roads, and housing in Appalachia were among the worst in the nation. To compound these problems, Appalachia lost the majority of its most productive people and most of its potential tax revenue through out-migration.[55] It is essential that future energy communities guard themselves from this destructive sequence of events.

ASSESSING AND PLANNING FOR IMPACTS

The problems we have described in this chapter can be minimized with planning and money. Communities should recognize that both proper planning and adequate funding for community needs will require active cooperation from the project sponsor. Some of the needed funds can be raised in the community, but most of the money must and should come from the project. To get help soon enough to be effective, the community must learn and be prepared to document the needs that the new project will impose upon the town. Citizens should insist that their government become informed, that it make the appropriate demands of a project sponsor, and that permission for the project be conditional upon meeting those demands.

In order to evaluate the impacts of the proposed facility, specific information is needed first on the project itself. A partial list of information needs is provided below.

1. Size of facility (in terms of fuel output) during experimental and construction phases.
2. Length and timing of construction period.
3. Planned operation start-up date.
4. Expected lifetime of project.
5. Number and skill level of construction workers.
6. Number and skill level of operation workers.
7. Transportation requirements associated with construction and operation of facility.
8. Ownership of facility and/or its site (federal involvement).

Such information is typically provided in an environmental impact statement (EIS), which must be prepared for many projects (see Chapter 8). If an EIS is not required for a project, the information should be obtained directly from the project sponsor, either by individuals or by officials of local government.

Once the basic parameters of the project are known, citizens can begin to assess the economic costs and benefits associated with it. As discussed above, a broad spectrum of services will be affected: housing, roads, sewers, medical facilities, schools, police and fire departments, and, last but not least, municipal government itself. An invaluable guide to assessing the expanding community's needs is *The Fiscal Impact Handbook: Estimating Local Costs and Revenues of Land Development.*[56]

FUNDING STRATEGY

Once a community has determined the new services that will be needed to support the proposed facility, it must evaluate the extent to which it can and is willing to bear these costs. Some of the needs, such as housing and medical services, will undoubtedly be provided in part by individuals in the community. However, local government should investigate whether private intentions will be adequate to meet the needs of the incoming population.

Other services, such as roads, schools, and police protection are more obviously the responsibility of the town.

A first reaction to increased revenue needs may be to raise property taxes. But, even if property taxes are increased to their legal limit, these revenues may not be enough. Moreover, such increases would impose unfair burdens on the elderly and those citizens unable to benefit from the new project. Federal involvement in the synfuels industry may impose additional tax burdens on local residents.[57] If an agency such as the Department of Energy (DOE) owns a synfuel facility and the land on which it is built, property tax revenues cannot be collected. Only if and when plant ownership reverts to private interests can property taxes be collected.

An alternative to raising taxes is issuing bonds, to be paid for by future tax revenues. Yet the bonds necessary to finance the essential community services can outlive the synfuels industry. Taxes for local residents may be forced even higher if the population dwindles after the closing of the synfuels plant.

A difficulty with the use of bonds to raise funds is that a city's legal bonding capacity may fall below its actual needs. For example, the city of Craig, Colorado, a coal boomtown, recently drew up a three-year plan, identifying the capital needed to accommodate its anticipated growth. Included in the budget were such essentials as water, sewer, drainage, street, and refuse collection improvements. The total capital construction needs for Craig totaled $30,200,000. The city, however, has a bonding capacity of $2,135,000.[58] Since in many states local bonding capacity is limited by constitutional and legislative provisions, it may be difficult for local governments to respond in an effective and timely manner.[59]

When local government is unable to come up with the increased funds to provide for necessary services, state governments may step in. States may assess additional property taxes or issue tax-free bonds. They may also disburse grants, loan guarantees or matching funds to assist localities hard-pressed by the costs of coping with development.

But make no mistake: it will be difficult to get the average state legislature to raise taxes, issue bonds, or give grants to help a community deal with the side-effects of "prosperity." A more likely source of state assistance is found in two existing sources of revenues available to most states, portions of which are supposedly earmarked for energy impact assistance: state severance taxes and state and federal mineral leasing revenues.

Severance taxes are imposed on the mining of a resource. Most states impose a severance tax on the mining of coal as well as other important energy and non-fuel minerals and oil and gas. In some western states, severance taxes may comprise a large portion of all state revenues (more than a quarter of all revenues in Wyoming and nearly 20% in New Mexico).[60] In many states, at least some of the severance tax revenues are distributed to local governments impacted by the mining which generated the revenue. In other states, particularly in the East, severance tax revenues are totally funneled into the general fund.

The availability of state severance fund revenues will be dependent on several factors: the size of the severance tax itself, the specific uses to which funds are dedicated, and the number of potential claimants.

Federal mineral leasing revenues are shared with the states in which mineral leasing takes place. Under Section 35 of the Mineral Leasing Act of 1920, states receive 50% of the revenues derived from sales, bonuses, rentals, and royalties from general activities on public lands within their boundaries. Revenues from the leasing and mining of federal coal and oil shale reserves are covered by Section 35, and by 1990 disbursements to the states should reach nearly $500 million per year.[61] Each state which receives federal mineral leasing revenues establishes its own formula for spending the money. The specific formulas vary, but almost every state makes provisions for putting some of these revenues back into energy-impacted communities. Many states also lease lands which they own for mineral development. This often provides an additional revenue source.

Severance taxes and mineral leasing revenues can play an important part in funding impact assistance in western states which generally have high severance taxes and contain large tracts of mineral-rich federal lands. A community affected by a synfuel plant in a western state will find its major problem to be the number of other communities—impacted by synfuels, coal mining, powerplant construction, expanded oil and gas exploration—also vying for the funds. The two funding sources are far less promising in Eastern states with low severance taxes and no federal lands.

The federal government currently has more than two dozen programs intended to help mitigate boomtown impacts; however, they are not well integrated and are somewhat ineffective. Generally, federal agencies have taken the position that leadership in this area must come from state and local authorities.[62]

The Powerplant and Industrial Fuel Use Act of 1978 specifically addresses the socioeconomic impacts of energy development. Designated "energy impact assistance areas" are eligible for planning development and site acquisition and development grants. These grants are not available to areas hit by oil shale development; they apply only to coal and uranium mining.

Communities seeking federal aid may encounter problems in the 1980s besides those caused by the fragmented federal program. Budget cuts advanced by the Reagan Administration have decreased the amount of available federal funding. Of the 35 federal financial assistance programs specifically aiding energy boomtowns listed in the U.S. Department of Housing and Urban Developmentls 1976 handbook, "Rapid Growth From Energy Projects: Ideas for State and Local Action,"[63] 23 had lower budgets for fiscal year 1981 than they had in fiscal year 1979.[64] Given the projected increases in U.S. energy development, the competition for funds from these federal programs should be significant. When tax revenues from the project, state funding, and federal aid are still insufficient to cover the costs to the community, you must rely on the project sponsor to provide needed assistance.

One technique suggested for mitigating boomtown impacts is to adjust project schedules. Extending the construction schedule for a coal gasification plant from five to eight years can reduce peak construction employment by 30%. This would reduce the discrepancy between the construction and operation workforce.[65] Likewise, if more than one plant is being built, the project schedules could be adjusted to minimize peaks and troughs in employment needs.

Success in dealing with the project sponsor hinges on being organized. The local government must prepare its own analysis of impacts and statement of needs. The community also should plan actively how and where it wants the new development to unfold. For example, it should consider questions such as whether a large shopping mall will provide desired convenience to community residents, or whether it will drive out existing local businesses. Finally, the local government must obtain a firm commitment from the project sponsor to make specified and timely contributions to the community's needs. A model for this process is provided by the growing group of communities that have succeeded in obtaining financial assistance from energy developers. Several of these cases are described below.

Rio Blanco County

The site of both coal and oil shale development, this western Colorado county obtained an agreement in 1981 from the Western Fuels Corporation to spend at least $15 million to help mitigate the impacts of a new coal mine. In Colorado, local government regulates all siting of energy facilities through land use permits. Rio Blanco County requires a project to obtain special use permits if it will employ a workforce equal to one-half percent of the county's population. The County Board of Commissioners used its control over these permits to get Western Fuels Corporation's cooperation in mitigating the impacts.

The board began by hiring a prestigious Washington law firm to represent the county. The county filed detailed comments on the project EIS, prepared by the Bureau of Land Management (BLM). Armed with the support of Robert Burford, Director of the BLM, the county began two weeks of negotiations with Western Fuels. In order to acccommodate the expected influx of 1,500 people associated with the mine (400 of them mineworkers), Western Fuels agreed to provide the following assistance:

—a $1.5 million cash block grant to the county;
—a $1.5 million cash block grant to Rangely (the town that will absorb most of the new population);
—a grant of $825,000 for new recreation facilities;
—a grant of $105,000 for library services;
—a grant of $876,000 for a hospital district;
—a grant of $120,000 for a fire disitrict;
—expenditure of $2.4 million on road construction;
—grant or loan guarantee of $2.26 million for new schools as needed;
—grant or loan guarantee of $3.75 million for a new water/sewage treatment system when needed;

—guaranteed housing for all the mine workers and "induced" workers (calculated by assuming an influx of 0.8 additional person for every member of mineworker family);
—payment of all additional municipal operating costs above and beyond ordinary inflation; and
—provision of $60,000 to pay someone to monitor the program.[66]

A key to Rio Blanco's success in negotiating this settlement was that the county first got the company to agree to the basic principle that the mine should result in no increase in taxes or user fees to the existing residents. Even the company's own studies then indicated that a multi-million-dollar expenditure would be needed to keep the promise.

Wheatland, Wyoming

As dicussed earlier in this chapter, Wheatland has had its share of problems from energy development in the past. When the Missouri Basin Power Project proposed a new coal-fired plant for the town in 1973, however, the community took action to prevent additional problems. In Wyoming, major new energy developments are under the jurisdiction of a state body called the Industrial Siting Council. By working with citizen groups in Wheatland, the Council was able to negotiate commitments for substantial amounts of money from the project (a consortium of municipal utilities and rural co-ops) to mitigate the impacts of the proosed 1500-MW Laramie River Power Plant. Success was due to the combination of the council's legal authority over the plant with the active involvement of Wheatland citizens in the planning process. Citizen groups were formed when the power plant was first proposed; and they began meeting on a monthly basis to discuss how the new plant would affect the town. The entire process took several years.

The package of assistance that the consortium agreed to provide demonstrates the versatility of approaches that can be used to help a town pay for the impacts of a large project. First, the consortium has undertaken a housing development for plant workers, including both bachelor and family accommodations. The consortium also provided $5.4 million to construct a new school, which the school district will repay when tax revenues from the project are realized. If the tax revenues from the new plant are insufficient to cover the costs, the district is excused from the obligation to repay the full amount. In addition, the consortium paid for the construction of an adult recreation center, and provided grants to cover the costs of fire, police, and mental health services. Finally, the consortium underwrote loans obtained from the state to pay for expanded sewer capacity; the loans will be repaid through water sales to customers.[67]

Evanston, Wyoming

Located in the energy-rich overthrust belt of the Rocky Mountains, this community did not obtain adequate financial assistance from energy developer until many of the problems described in this chapter had already begun. Much of the development in the area was initiated before the Industrial Siting Council was established, and what little planning for impact mitigation that occurred was done by the companies themselves. When these efforts proved to be entirely inadequate, the companies commissioned a study which subsequently demonstrated the need for substantially increased mitigation assistance. As a result, the companies are contributing a total of $2 million for various improvements in the town of Evanston.

The main impetus for the original mitigation effort was the planned construction of two processing plants by Amoco and Chevron, the two major developers in the area, to begin in the spring of 1980. The companies decided to hire only single male construction workers, and built bachelor camps to house the estimated total peak workforce of 1000. The companies also each donated $500,000 to local governments in Uinta County to pay for additional services required as a result of the population influx. However, the construction workforce has grown to twice the originally projected level, and the initial $1 million contribution was obviously not enough to mitigate all the problems.

Recognizing that a much larger effort was needed, the two companies joined with other energy developers in the area to form the Overthrust Industrial Association, which in turn recruited a study of the existing and projected future socioeconomic impacts of energy development in the area. Following the completion of the study, citizen task forces were established to identify specific needs. As a result of these efforts, the association has agreed to provide assistance including the following items:

—a grant of $500,000 toward the $3.5 million cost of building a new-county courthouse;

—a contribution of $300,000 to replace burned out pumps at the overloaded Evanston sewage treatment plant;
—a grant of $40,000 for new parks;
—a grant of $15,000 for an arts council;
—a promise to fund half the salaries of a "human services coordinator" and an alcohol rehabilitation counselor for three years;
—an interest-free loan of $150,000 to construct a child care center;
—a grant of $140,000 to develop a "master plan" for projected Evanston growth; and
—a grant of $95,000 to fund a city/county survey of growth patterns.[68]

The East

While this chapter has focused on boomtowns in rural western areas, synfuels development in the East will also have adverse impacts on the social and economic stability of eastern towns. In most cases, eastern towns are closer to urban centers than their western counterparts; some workers will commute to the plant. This reduces the number of new residents in a town and the associated shortages of housing, school, and medical facilities, etc.[69] According to the DOE rating scale mentioned earlier in this section, most of the counties in the Midwest and Appalachian regions have a "medium high" assimilative capacity. A few counties in each region are in the "medium low" or "high" categories.[70] Despite these factors, eastern synfuel development areas face two problems that western areas do not.

First, many eastern towns have already been through the boom and bus cycle because of traditional coal mining. With synfuels development, they will have to face the cycle again, rejuvenating deteriorating facilities, calling back the youth who abandoned the stagnating economies of the past few years, yet all the while knowing that the bust is soon to follow. This results in a "seesaw" economy.

The second problem unique to eastern synfuel communities is the likelihood of several synfuel plants being developed near each other. Each facility represents only a small part of the total energy and industry related growth anticipated over the next decades.[71] The combined impacts of several projects could cause problems of the same magnitude as those faced by western regions.

7

Citizens, Synfuels and the Environmental Laws

Air, water, and land are resources. We use them in a variety of ways—to fulfill biological needs, for recreation, for agricultural and industrial production, and to dispose of wastes from all of our other uses. Some of our uses of the air, land, and water are incompatible with other uses. Land on which an industrial plant is built cannot be used for agriculture or recreation. Water that is polluted cannot be drunk.

Americans have developed a variety of ways to decide how the air, land, and water may be used, all expressed in the form of legal rules. Land is owned. The land owner largely controls its use, although limits may be imposed by both zoning laws and the many other laws that restrict human activity. (E.g., you may not grow marijuana on your own land.) In many western states the right to use a specific portion of the flow of a river or stream may be owned. The air, on the other hand, and most water, is not "owned," but rather used on a first-come, first-served basis. Until recently that has meant that the air and water have been used freely as cesspools into which we dump our wastes—the by-products of everything from automobiles and woodstoves to powerplants and papermills.

During the 1960s the nation increasingly became aware that pollution—the uncontrolled disposal of wastes into the environment—was not cost-free. We learned that pollution damaged health and property. We found we had lost the use of resources that were precious to us. People demanded that something be done, and Congress responded by passing a collection of laws designed to protect our common resources—the environmental laws. Many will have an impact on synthetic fuels plants.

A dozen years ago, Garrett Hardin wrote about the destruction of village commons in 18th century England as a result of overgrazing, the inevitable result of the fact that each herdsman benefitted individually from adding to his herd, while the consequences fell upon all users of the commons. This tragedy, he noted, reappeared in the problems of pollution. "Here it is not a question of taking something out of the commons, but of putting something in—sewage, or chemical, radioactive, and heat wastes into water; noxious and dangerous fumes into the air . . . ."

First we abandoned the commons in food gathering, enclosing farm land and restricting pastures and hunting and fishing areas. These restrictions are still not complete throughout the world. Somewhat later we saw that the commons as a place for waste diposal would also have to be abandoned. Restrictions on the disposal of domestic sewage are widely accepted in the Western world; we are still struggling to close the commons to pollution by automobiles, factories, insecticide sprayers, fertilizing operations, and atomic energy installations.

Hardin concluded with the observation:

Every new enclosure of the commons involves the infringement of somebody's personal liberty. Infringements made in the distant past are accepted because no contemporary complains of a loss. It is the newly proposed infringements that we vigorously oppose; cries of "rights" and "freedom" fill the air. But what does "freedom" mean? When men mutially agreed to pass laws against robbing, mankind became more free, not less so. Individuals locked into the logic of the commons are free only to bring on universal ruin.

—*Science,* 13 December 1968, 1245, 1248.

Most of the environmental laws require federal agencies (sometimes in cooperation with state agencies) to establish and enforce rules that apply to potential polluters. For example,

Federal Environmental Laws that May Apply to Synfuels Plants

1969	National Environmental Policy Act (NEPA)	Requires Federal officials to analyze and consider the environmental consequences of their actions. Requires written environmental impact statement on major actions.
1970	Occupational Safety and Health Act (OSHA)	Requires safe workplace. Authorizes Department of Labor to set and enforce standards.
1970	Clean Air Act	Requires EPA limits on air pollution from industrial sources and automobiles.
1972	Clean Water Act	Requires EPA limits on pollution from industrial and minicipal sources.
1973	Endangered Species Act	Requires Department of Interior to designate endangered species and forbids federal action that would injure such species.
1974	Safe Drinking Water Act	Requires EPA to set standards to protect drinking water.
1976	Resource Conservation and Recovery Act (RCRA)	Requires EPA to set standards for transportation and disposal of hazardous wastes.
1976	Federal Land Policy and Management Act (FLPMA)	Requires the Department of Interior to plan for and manage use of lands belonging to the United States to permit multiple uses, over sustained periods in an environmentally sound manner.
1976	Toxic Substances Control Act (TSCA)	Requires EPA to monitor the manufacture of chemical substances and act to prevent threats to human health and safety.
1977	Surface Mining Control and Reclamation Act (SMCRA)	Requires Office of Surface Mining in Department of Interior to set strict standards to prevent environmental damage from surface or underground coal mining.
1980	Comprehensive Environmental Response, Compensation and Liability Act (Superfund)	Sets up fund administered by EPA to clean up toxic spills and abandoned dump sites.

EPA has set rules limiting the amount of sulfur different types of industry may release into the atmosphere, and the Office of Surface Mining has set rules for how land must be restored after it is mined.

Some of the environmental laws require large new industrial facilities to obtain permits before construction begins or before the plant begins to operate. The sponsor of the plant must submit information showing that the plant will conform to the law to the satisfaction of the government agency responsible for administering the law in question. The agency, if it is satisfied that the plant will comply with the law, issues a permit. The permit is often a complex document that details specific conditions that the sponsor must meet in the design, construction, and operation of the plant. Obviously the conditions imposed in a permit have a significant effect on how much pollution comes out of a plant. Denial of a permit (a rare occurrence) prevents construction of a plant.

This chapter discusses in some detail the permitting process under the Clean Water Act and the Clean Air Act, and explains relevant provisions of nine other federal environmental statutes.

Of course, it is not only federal laws that will apply to synthetic fuels plants. Every state, and most county and municipal governments exercise some sort of permitting authority over major new industrial facilities. Historically, state and local governments often have done more to assist and encourage the builders of new facilities—in order to obtain new jobs and tax revenues—than to control the manner in which these facilities are built or operated. One reason for the passage of federal environmental laws was the pressure on individual states to offer lax laws—a "favorable regulatory climate"—to attract industry.

Despite past problems, some states and localities now approach proposed industrial development with intelligence and care. State and local proceedings may provide excellent opportunities for citizen participation, especially if local concern is widespread and organized enough to be politically significant.

Because the decision whether to grant a permit rests with a government agency acting pursuant to specific statutory standards, it provides an important opportunity for citizens to influence the design and operation of a synfuels plant. Most laws explicitly require that the public be given an opportunity to participate in permitting proceedings. If the permitting agency does not apply the law properly, citizens *may* be able to challenge the permit in court. Permit challenges *based on valid objections* can cause long delays in plant construction. Such delays are extremely expensive for plant sponsors. As a consequence, they may be particularly responsive to relevant and well-grounded objections raised before and during the permitting process.

The permitting process also provides an opportunity to get information about a proposed plant. The sponsor of the project must submit a great deal of information with his permit application to show that the plant will meet relevant legal requirements.

Finally, the permitting process, which often involves public hearings, provides a focal point around which community concerns may be organized. That is most likely to be true if the issuance of the permit involves decisions of genuine concern to the community.

Permitting proceedings, and the laws that require them, are complicated and technical. Participation requires time, hard work, and often scientific or legal assistance. Not all environmental hazards are subject to any legal standards. For example, the section on air pollution describes the federal Clean Air Act, under which EPA regulates 11 pollutants. Many substances known or suspected to be dangerous to human beings that may be emitted by synfuels plants are not yet regulated under the Clean Air Act. Nor do all environmental statutes require permits. The Toxic Substances Control Act, for example, only requires the submission of limited information to EPA in certain circumstances.

Many laws leave wide discretion to the permitting agency to decide on the appropriate terms for a permit. The pressures on the agency to act quickly and set terms consistent with those sought by the applicant may be very great. The present Administration has seemed most eager to use its discretion to weaken environmental requirements.

Citizens should make a careful decision whether they want to put their time and resources into participating in permitting procedures. If the answer is yes, the following simple rules may be helpful.

1. *Start Early.* It takes time to gather information and learn the ropes. Also, long before the legal deadlines for participation, federal and state officials begin negotiating with project sponsors and forming their own opinions.

2. *Get As Much Information As Possible.* The more information you have about the plant, the site, the plant sponsors and the permitting process, the more effective you will be. Talk to everyone. federal, state and local officials, the project sponsors, university scientists, the local medical society; anyone who might know about the issues involved. Do not argue, just take notes.

3. *Do Not Assume Anything.* If you see something that does not make sense, ask about it. Do not assume "they could not make a mistake like that," or "I must be stupid, they must know what they are doing," or, worst of all, "if it is a problem, the Government will deal with it." The Diablo Canyon nuclear power plant was licensed and about to begin operation before it was discovered that the engineers building it got the blueprints confused and built one major safety system backwards.

4. *Be Accurate.* Citizens start with a handicap. The fact that they want to participate seems to make people think they are a little crazy. So you have to build your credibility in order to be taken seriously. Rhetoric and exaggeration do not help; they hurt in the dull gray confines of the permitting process.

5. *Do Not Forget Politics.* Permitting is governmental action. All governmental action is, at least at some level, political. Being knowledgeable and credible is very important, but being organized and numerous helps a lot too.

We will be dealing with a series of laws or statutes which have been passed by the U.S. Congress. These laws may be referred to in any of three ways. By common name, e.g. the Clean Air Act. Laws are also referred to by public law number, e.g. P.L. 95-95 (also the Clean Air Act). P.L. stands for "public law," the first number tells you which Congress passed the law, in this case the 95th Congress, and the second number tells which bill, in numerical order, it was, in this case the 95th bill signed into law during that session). Laws are also referred to by U.S. Code number, 42 U.S.C. 1857 (again the Clean Air Act. The first number tells which volume of the U.S. Code it is in, while the second number tells you the section in that volume).

The U.S. Code can be found in any library in America. But the section numbers used in the Code are rarely those found in discussions of the law. Usually people refer to the law as it appears in its public law or "slip" form.

Fortunately, the U.S. Code provides the public law section numbers in parentheses at the bottom of each paragraph. So, if the document you are reading seems to have nothing at all to do with what it should, check to see that you are not looking for U.S. Code numbers in a public law print or vice versa.

If you read the text of one of the environmental statutes, notice how little it says about pollution. Most of the major environmental laws tell the administering agency, normally the Environmental Protection Agency, how to establish specific rules and regulations governing emissions levels and ther permitting process. Federal rules and regulations are published in two places.

The first is the Federal Register, a daily publication containing all of the rules, proposed rules and notices issued by all federal agencies. The Federal Register is in many libraries. When you see a reference to a document in the Federal Register, it will often look like this: 45 FR 33063 (May 19, 1980). The "45" refers to the Volume number, 45 is 1980, 46 was 1981, 47 is 1982 and 48 will bc 1983. "33063" is the exact page number.

A daily publication in which today's rule from EPA is printed next to today's meeting notice from the Export-Import Bank is handy for keeping abreast of changing developments, but it hardly provides an overview of all rules that might apply to a given permit. For that reason, all currently valid agency rules are put in the Code of Federal Regulations (CFR), a massive compendium that runs over one hundred volumes. Normally, all of the regulations issued by an agency will be found in one volume. All of EPA's regulations are found in 40 CFR (which while it is called one volume is so large that, this year, it is packaged in nine separate books); Department of Energy rules are at 10 CFR; rules governing the National Park System at 36 CFR; Fish and Wildlife at 50 CFR; minerals at 30 CFR and public lands at 43 CFR.

The Code of Federal Regulations contains only the language of regulations plus some important appendices. When regulations are first printed in the Federal Register, they often include lengthy preambles that explain the rule. Even if you have a copy of a regulation in CFR, you may wish to look at the issue of the Federal Register in which the regulation first appeared (it will be listed in the CFR presentation as "Source").

An important note: The Code of Federal Regulations is compiled once a year for each agency. (The volume shows the date.) All final regulations promulgated before that date are contained in the volume, but of course not changes after that date. To see if there have been changes you must ask the agency or look in the Federal Register at the periodic Guide to CFR Parts Affected (a daily version is found at the front of each issue and a cumulative monthly list is in the back), that references all changes since the most recent CFR publication.

It is not as difficult as it sounds. And if you ever have any trouble, call the Federal Register. The number can be found in the Reader Aids section at the back of each issue.

We have used some technical terms to describe rules and regulations in this discussion. An agency can not just come up with a rule and print it, saying "This is it folks." Generally, a copy of a proposed rule is published in the Federal Register and citizens and industry are given an opportunity to comment on the rule. (For very complicated rules, sometimes an Advance Notice of Proposed Rulemaking is published first.) The proposed rule is written in precisely the format that the final rule will appear in. After the comment period is over and the agency makes changes, if any, in response to the comments, the agency then publishes or promulgates a final rule. Proposed rules and final rules are printed in different parts of the Federal Register. Only final rules count.

THE CLEAN AIR ACT

The Clean Air Act is probably the most visible and best known of the environmental laws passed in the last decade. It has established a complex system designed to move the nation slowly towards healthy air, while allowing for continued growth in areas where the air remains clean, even in areas which currently have dirty air.

One reason why disputes over clean air are so difficult and contentious is because of arguments over the relation of emissions from a plant to pollution levels in a surrounding area. Most of the EPA and state regulations discussed below establish emission limitations in order to reduce concentrations of a pollutant in the atmosphere, rather than to reduce tonnage or output over time of a plant. While gross output from the plant is one variable in determining ambient air concentrations of a pollutant, terrain, wind patterns, atmospheric conditions, stack heights, and weather all play important roles.

Generally we are forced to estimate the air quality impact of a specific plant's emissions by computer modeling. There are two steps in doing this. First, there is a need to monitor the air at the proposed site to obtain basic information on existing pollution concentrations, wind patterns, and meteorological data. This background data is then used in an EPA-approved computer model to estimate just what impacts a proposed plant will have.

This often turns clean air permitting exercises into battles between computers, very expensive and often incomprehensible to the layman.

Criteria Pollutants

As noted earlier, the Clean Air Act does not regulate all potentially harmful air pollutants. Following Section 109 of the Clean Air Act, the Environmental Protection Agency has established national ambient air quality standards (NAAQS) for six major air pollutants: carbon monoxide, sulfur dioxide, nitrogen oxides, particulate matter, ozone, and lead. (A standard has also been established for nonmethane hydrocarbons, but it is used only as a guide to assist in meeting the ozone standard). Air pollution control under the act is almost entirely focused on the regulation of these seven "criteria" pollutants—so-called because EPA develops "criteria documents" on their health and welfare effects before promulgating standards.

There are two kinds of ambient air quality standards: *Primary standards* establish the level of air quality necessary to protect human health—even the health of sensitive people such as asthmatics—with an adequate margin of safety. *Secondary standards* (which exist for only some of the criteria pollutants) establish levels necessary to protect "public welfare," which includes plant and animal life, visibility, buildings, and materials.

These standards do not by themselves apply to specific industrial plants; they apply to the air in broad areas called air quality control regions (ACQRs). Based on the levels of pollutants in a region, an ACQR or a portion of it may be classified as meeting air quality standards (attainment), exceeding them (non-attainment), or unclassified. The status of an ACQR plays a key role in determining the type of air quality regulation which might be applied to a new synfuels plant. It is important to note that a single ACQR might be in attainment status for one pollutant, non-attainment for another, and unclassified for a third; there is not a single rating.

Hazardous Air Pollutants

Most regulation under the Clean Air Act has involved control of the seven criteria pollutants. As the Council on Environmental Quality stated in its 1980 annual report: "To date, the criteria pollutants have been the major focus of air pollution control efforts. But increasingly more attention will be devoted to control of hazardous air pollutants under Section 112 of the Clean Air Act."

Under Section 112, the Administrator of EPA is directed to list substances to which no ambient air quality standard applies but which "in the judgment of the Administrator causes, or contributes to, air pollution which may reasonably be anticipated to result in an increase in mortality or an increase in serious, irreversible, or incapacitating reversible, illness." Seven substances have been listed as hazardous under Section 112: asbestos, mercury, beryllium, vinyl chloride, benzene, radionuclides, and arsenic. EPA has issued final emissions standards for four of them: asbestos, beryllium, mercury, and vinyl chloride. It is also considering listing other pollutants including polycyclic organic matter and cadmium.

Unlike criteria pollutants, regulations governing hazardous pollutants apply only to new facilities. EPA issued a proposed methodology in October 1978 in which it indicated an intention to use Section 112 as a way of regulating the airborne emission of carcinogens. EPA has not yet adopted this approach in final form.

State Implementation Plans (SIPs)

The authority to establish criteria standards lies totally with the federal EPA, but the basic authority for seeing that areas meet the standards lies with the states. Section 110 of the act requires each state to submit an implementation plan to set forth the steps that it will take to meet the primary and secondary standards for each pollutant subject to an NAAQS. The plan must show that the primary standards will be met within the statutory time limit and the secondary standards will be met within a reasonable time thereafter. A State Implementation Plan (SIP) must be approved by EPA. EPA can approve the state plan, only if the state has the authority to:

1. Adopt emission standards and limitations.
2. Enforce laws and regulations and seek injunctive relief.
3. Reduce pollution emissions on an emergency basis.
4. Prohibit the construction or operation of any source that will prevent either the attainment or the maintenance of any air quality standard or that will interfere with efforts to prevent significant deterioration.
5. Obtain all necessary information.
6. Require owners and operators of stationary sources to install emission monitoring devices.
7. Demonstrate that all standards will be maintained within the statutory time limits.

The states are responsible for formulating implementation plans and making whatever policy decisions they deem appropriate in terms of allocating costs and responsibilities for pollution cleanup, so long as their final approach will bring all parts of the state into attainment.

If the Administrator of EPA deems the state plan inadequate, he has the authority to promulgate his own plan. This has been done with only one state, Ohio. In practice, EPA normally approves a plan "with exceptions" and issues federal standards and enforcement procedures to meet specific deficiencies. In doing so, the Administrator is limited to using authorities granted by the Clean Air Act. In other words, if the Administrator disapproves of such elements as numbers, dates, or procedures, he can change them. But if he finds that a state agency lacks a particular required authority or even that no state agency exists, he cannot vest the agency with the missing power or create a new agency. He can, however, return or leave the regulatory function to the federal EPA. This has happened with many of the important permits we will discuss below.

So, in every state most clean air regulation is in state hands. Some state agencies have assumed full responsibility. The states have great leeway to determine how to clean up *existing* sources of pollution, providing their approach seems reasonably likely to meet the NAAQS. The states are also free to implement *stricter* requirements if they see fit, and some have done so. Thus, the initial focus of clean air decisions is likely to be at the state level *unless* issues are raised involving matters for which the SIP has not been approved. As we shall see, the most important permit requirement facing a synfuel plant often falls into this category.

New Source Regulation

All synfuel plants are "new sources" under the definitions found in the Clean Air Act, and are subject to rules specifically designed to regulate pollution from new sources.

When a new industrial facility is constructed in a complying (either attainment or unclassified) ACQR, it must demonstrate that its emissions will not result in non-attainment of any of the NAAQS. The control of specific permit conditions which might be applied to a plant in this circumstance lies almost completely with the state in which the plant is located. That state must have developed an EPA-approved State Implementation Plan which indicates must how it will establish and maintain the NAAQS for each pollutant. A SIP would not have received EPA approval if it allowed a new source to be constructed in a way which would violate an NAAQS. But within these broad limits, a state is largely free to determine what conditions it does or does not want to impose on a new source in an *attainment area*. Thus, the limits are those found in state law.

But for certain large, technologically complex new sources of pollution, Congress has taken away much of that freedom to determine how to maintain air quality. There are four important programs mandated by the Clean air Act which may appply to the construction of new sources. In each case, the Act requires a *preconstruction review* of the new source and the application of some technical standards.

Section 111 of the Clean Air Act directs EPA to establish *standards of performance* for new sources, if the source "causes or contributes significantly to our pollution which may reasonably be anticipated to endanger public health or welfare ..." These *new source performance standards* (NSPS) are designed to promote uniformity and prevent bidding wars between States for the location of new industry.

Plants subject to NSPS are required to install pollution control equipment which utilizes the *best available control technology* (BACT) for continuously reducing emissions. The specific requirements are codified in Part 60 of Title 40 of the Code of Federal Regulations. The NSPS requirements generally apply only to one or two pollutants emitted in large amounts by any particular type of source. In establishing standards of performance, EPA has also regulated the emissions of three noncriteria pollutants from specific new sources: sulfuric acid mist, fluorides, and total reduced sulfur.

EPA has established standards for 30 categories of sources and has published a list of 59 proposed major sources for future coverage. Synthetic fuel technologies are not on that list, so standards of performance are not likely to be developed for years.

The absence of a new source performance standard, however, does not return control over synfuel plant construction to general state standards, because almost any plant will be covered by the preconstruction review provisions under the non-attainment or prevention of significant deterioration (PSD) programs. Any new major source (commercial-sized synfuel plants will all meet the minimum pollution criteria for being major) locating in a non-attainment region must undergo a *non-attainment review* to determine that it does not contribute to an increase in atmospheric loadings of the pollutant for which the region is not in compliance. Any major source locating in a region which is in attainment or which is unclassified for any pollutant must undergo a *PSD review* which will prevent significant deterioration of the atmosphere for *any* pollutant which the source emits in excess of a minimum amount established by EPA.

The two reviews are not exclusive: a plant may locate in an area which is non-attainment for one or more pollutants and attainment or unclassified for others and be required to meet the requirements of either or both reviews, depending on the pollutants it emits. Therefore, in practice, every new synfuel plant will be subject to a set of air-specific air quality requirements.

Non-Attainment

If the pollution in an area exceeds any NAAQS, it is considered a "nonattainment area" for that pollutant. Proposed facilities in nonattainment areas are exempt from PSD review for these nonattainment pollutants. Instead, the proposed facility must meet stricter requirements.

A new plant which would locate in a non-attainment area, of course, faces serious siting problems. By definition, any new pollution just makes an already unhealthy situation worse. However, so as not to strangle economically non-attainment areas, the Clean Air Act provides for a method by which plants may be located in non-attainment areas. It places stringent conditions on the project.

In order to locate in a non-attainment area or to build a facility which would affect a non-attainment area, four conditions must be met:

1. *Offset.* The new source must obtain emissions "offsets" by cleaning up existing sources, either by cutting pollution controlled by other polluters or buying out and closing polluters, in order to offset its own new emissions. The offsets must at least equal the projected new emissions of any pollutant for which the area is not in attainment.

2. *Compliance.* The owner of the source must show that any other facilities it might own in that state are in compliance with all clean air regulations. If it is violation at another source, it must demonstrate that it is following its compliance schedule to reduce pollution there.

3. *Technology.* The proposed plant must achieve the lowest achievable emission rate (LAER) for any non-attainment pollutant. LAER is the more stringent of (a) the most stringent emission limitation required by any state for this kind of source, or (b) the most stringent emission limitation achieved in practice. LAER is very strict; it permits only a very limited consideration of costs. In most cases to date, LAER has proven to be equivalent to the NSPS for that source. For synthetic fuel facilities, because there is neither an NSPS nor existing plants, it is difficult to forecast what impact the LAER requirement will have. Some state rules, however, require that technology from similar kinds of processes be examined as a basis for LAER decisions. This is called "technology transfer."

4. *Positive Net Air Quality Benefits.* There must be a net reduction of pollution in the area from the time the permit application is filed to the time operation begins. Although air quality improvement is not required at every location affected by the source, the overall air quality in the same general area must be improved.

If the state SIP does not contain adequate provisions to bring non-attainment areas into compliance within deadlines established by the act, EPA is required to impose a *construction ban* on all new construction. Such a ban would obviously prevent construction of any synfuel plant while it is in effect.

Visibility

The Clean Air Act contains a section designed to protect visibility in National Parks over 6,000 acres and wilderness areas over 5,000 acres created before 1978 from degradation by new and existing sources.

Some measure of visibility protection is already built into the PSD program, which requires an analysis of impacts on visibility as a part of preconstruction review (described in next section). Under the PSD program, a state is required to deny a permit, if it can be shown that a new source would have an adverse impact on visibility within a Class I area. A state also has the authority to require a new source to analyze different sites, install additional control technology, or limit its operating conditions if these are necessary to mitigate adverse impact.

EPA's first major rulemaking under Section 169 made a few changes in the program as it would apply to synfuel plants. Like the PSD program, the visibility program will be incorporated—after EPA approval—into state SIPs, leaving enforcement and decisionmaking with approved states. Two new powers stemming from Section 169 add requirements not present in the current PSD program. First, states will be required to analyze the visibility impact of sources not now covered by the PSD program. Sources locating in a non-attainment area which might have an impact on visibility in a Class I area will be examined; they would not be under the PSD program.

Second, the visibility regulations allow for protection of visibility in areas outside a Class I area, if the area is part of an *integral vista.* Integral vistas are "important views from a point in a mandatory Class I area of a scenic viewpoint outside of the area." Overlooks are the most prominent example. The identification of a large number of integral vistas, of course, would expand dramatically the impact of the visibility program. States, however, are not required to protect integral vistas until they have been formally designated by the federal land manager of the Class I area. In most cases the National Park Service is the land manager, but Class I wilderness aras may be under the management of the Forest Service, the Fish and Wildlife Service, or the Bureau of Land Management if the wilderness is located in a national forest, a wildlife refuge, or on public lands, respectively.

While proposed guidelines have been published, no final formal designations have been made. Unless an integral vista has been designated six months before the revision of the relevant state SIP, the state need not consider protection of that vista in performing its permit review under that SIP. Nor must a new source analyze visibility impacts on an integral vista unless the vista was identified six months before the permit application was filed. Under EPA regulations, a state can decide not to protect an integral vista, if it concludes that other factors such as energy or the economy are more important. Therefore, at present, there are no integral vistas, nor will the protection of integral vistas apply to any synfuel facility which has already filed for an air permit.

Prevention of Significant Deterioration

Areas of the country with air quality that is better than the National Ambient Air Quality Standards for one or more pollutants are designated "attainment" and are subject to the Prevention of Significant Deterioration (PSD) provisions of the Clean Air Act.

This is a long-term program for the conservation and wise use of our clean air resources. It helps protect pristine air and visibility in valuable natural areas and prevents air in other places from deteriorating to the polluted levels of major cities and industrialized zones.

The PSD program is implemented through a *permit* requirement. Major new sources that want to locate in clean air areas and existing sources that want to expand must demonstrate that planned emissions from the plant will not exceed applicable requirements. The permits are issued by EPA or by states which have EPA-approved State Implementation Plans (SIPs) or that have been delegated PSD-permitting authority.

Every commercial-scale synfuels plant will require a PSD permit.

Although the prevention of significant deterioration program of the Clean Air Act is expressly directed towards minimizing increases of two criteria air pollutants in clean air areas, the PSD permitting process can be used to apply some measure of control to any federally regulated air pollutant—even if EPA has not established ambient air quality standards. In other words, the PSD permit process can be used to control—to the maximum extent possible—emissions of any of the 15 air pollutants that EPA regulates under various Clean Air Act provisions.

The fact that a permit is required provides an important opportunity for citizens to influence the design and operation of a proposed synfuels plant and even where, or whether, it is built. Citizen leverage is at a maximum during a permit proceeding. The developer cannot go forward until the permit is granted. The burden of proof to show that applicable standards will be met falls on the party seeking the permit. And federal and state agencies are required to review the complex factual submissions and analyze them at their expense.

Concerned citizens should consider participation in the PSD permit proceedings. But PSD permitting is technical and complex. Effective participation requires hard work and probably some help from experts.

The Clean Air Act is designed not only to clean up areas of the country with dirty air but also to prevent "significant deterioration" of air quality in the areas with clean air—those classified as in "attainment" status or unclassified for at least one of the seven criteria pollutants.

Under the PSD program clean air areas are designated in three classes:

Class I areas are wilderness areas greater than 5,000 acres and national parks greater than 6,000 acres which were in existence in 1978. New wilderness areas created after 1978 do not automatically become Class I areas. States may designate additional areas as Class I.

Class II initially included all other clean air regions. States may designate any Class II lands as Class I; they may also designate Class II lands as Class III, except that the following may not be designated Class III: national monuments, primitive areas, preserves, recreation areas, wild and scenic rivers, wildlife refuges, lakeshores or seashores that exceed 10,000 acres. Post-1978 parks or wilderness areas which exceed 10,000 acres may not be designated Class III either.

Class III areas are those designated by a state, pursuant to procedures in Section 164. No areas have been designated Class III.

The PSD program is based on a system of budgets ("increments" in the language of the act) that specify the allowable increases in pollution in clean air areas. The Clean Air Act specifies increments for two pollutants that stem mainly from stationary sources—sulfur dioxide and particulates. There are different sized increments for Class I, Class II, and Class III areas; Class I has the smallest increments, Class III the largest.

Although there are NAAQS for all seven criteria pollutants, PSD increments are measured only for particulates and sulfur dioxide. Thus, the upper limit for pollution in the area is the baseline plus the increment (the "ceiling") or the NAAQS, whichever is less. In no case may ambient pollution levels for the seven criteria pollutants be permitted to rise above the NAAQS.

Congress intended the states to implement the PSD program under EPA supervision, but the states must develop an adequate PSD program before EPA delegates PSD authority to them and few have as yet. EPA has delegated some of the authority for aspects of the program (particularly receipt of applications and technical review) to some other states, and for new sources in some states the program is administered by EPA alone.

PSD Permitting: Step by Step

The following pages outline the PSD permit process. Where EPA administers the entire program, the standards for the program will be those found in the Code of Federal Regulations at 40 CFR 52.21—17 pages of material—and the procedures for making a decision on a permit application, holding hearings and the like will be found in EPA's rules for its Consolidated Permit Program at 40 CFR 124. Permit applications will be made to an EPA Regional Office, all review and decisions will be made by EPA personnel, and the decision can be appealed to the Administrator.

Where a state administers the entire program, its rules can be found in the State Implementation Plan. That plan must meet the requirements set forth at 40 CFR 51.24. When a state administers the PSD program, applications are made to the state, procedures are governed by state law, and administrative appeals take place within the state administrative structure.

Where states have been delegated some authority, it is necessary to find out—on a state-by-state basis—just what that delegetion is. In most cases, delegated states receive the applications and perform completeness and technical reviews, but final decisions or, at least, appeals from final decisions remain with the federal EPA. But, if you live in a state with some delegated authority, check with both the state air quality agency and the EPA regional office to find out just what rules govern that delegation.

Check with the EPA office for your region of the country and with your state air quality office. The following description is of the process as administered by EPA.

A PSD permit is required before construction of any new or modified "major stationary source." If a synfuels plant will emit more than 100 tons per year of any regulated pollutant, even with pollution controls, it is considered a "major" source. The 100 tons includes both emissions from stacks and "fugitive" emissions, such as leaks. However, EPA has proposed to ignore fugitive emissions, as part of a settlement of an industry lawsuit.

The permit must be obtained before construction begins (unlike the NPDES water pollution permit, which does not need to be acquired until operation begins). "Construction" does not include preliminary site preparation but is considered to commence when concrete is poured or substantial contracts are signed. If construction begins before the PSD permit is granted, section 167 of the Clean Air Act states that the EPA Administrator or the states can seek injunctive relief. Citizens can too: section 304(a) (3) states that "any person may commence civil action . . . against any person who proposes to construct . . . without a permit . . ."

A PSD permit expires if construction does not begin within 18 months after it is issued, but this can be extended.

The formal permitting process begins when a company submits an application. However, prior to filing an application, companies often have pre-application conferences, with the permitting agency. These meetings are important, for they often establish the ground rules for the technical and analytical content of the application. No public notice of these conferences is required by federal law. However, you can request your agency to let you know of such meetings. Some agencies will cooperate. (A possible source of this information may also be the Federal Land Manager of any affected lands— national parks and forests, BLM lands, etc.—who is often informed in advance of applications. Check with them too.) Finding out can be very helpful, because much of the hard analysis will take place before the public formally is involved. This creates two problems. First, there is a good deal of interaction between the developer and the permitting agency before any other views are fed into the process. Second, the effective response time for citizens is shortened: the developer may have two years to put together his application and the agency six months to analyze it, while the citizen is given 30 days to respond. You can lessen these drawbacks by getting yourself into the process before the formal public participation segment begins. Ask to meet with EPA or the state, scour the literature for information on the process, use the Freedom of Information Act (and state counterparts) to get information. Make them deal with you before the draft permit is issued.

Once the permit application has been submitted, EPA must notify the applicant of any deficiencies in the application within 30 days.

If the permitting agency certifies an application as complete when in fact it lacks important information, or if it has used improper models, or does not consider important alternatives, little can be done to reverse the process and insure that the application is truly complete. And formal citizen involvement is especially difficult at this stage. To participate effectively, you must get the application, perform an analysis, determine what was wrong with the application and notify the permitting authority—all within the 30-day limit; all the more reason to get involved early, well before the completeness determination is to be made, and have your concerns considered as the application is being filed and its completeness determined.

Formal public participation does not begin until EPA, based a technical review of the application, makes a preliminary decision on the application.

EPA then provides public notice of its preliminary decisions by mail to persons who have placed their names on a mailing list and by publication in a daily or weekly newspaper in the affected area. Those who do not read the small print legal sections of weekly newspapers for enjoyment will want to be on the mailing list. (There is no requirement that notice of draft PSD permits appear in the Federal Register.) After that, the agency must provide an opportunity for written comment, and a hearing if requested.

c. *What goes in the permit application?*

The permit application must contain data and analysis to show that the plant, as proposed, will comply with Clean Air Act requirements. Those requirements, once again are that:

—the plant utilize *Best Available Control Technology* (BACT);
—the plant meet applicable *New Source Performance Standards* (NSPS);

—the emissions from the plant not exceed available *Prevention of Significant Deterioration* (PSD) "increments" for particulates and sulfur dioxide;
—the emissions from the plant not cause a violation of the *National Ambient Air Quality Standards* (NAAQS);
—the emissions from the plant not exceed applicable *National Emission Standards for Hazardous Air Pollutants* (NESHAP).

To meet this burden, the applicant must submit:

—*Process design information* showing pollution control measures and expected levels of emissions;
—*An analysis of potential impacts* from the plant on other than air quality;
—*Ambient air quality monitoring data* showing the existing level of pollution for all regulated pollutants that will be emitted from the plant;
—*Computer modeling* to show that the plant will not cause volations of applicable standards.

We will review those components in order, but we emphasize that the purpose of the application is to demonstrate that the plant, as yet unbuilt, will meet clean air standards. The complexity of the process is a consequence of the necessity of predicting from a design what will happen in the real world. Once the plant is built, it is expensive and politically difficult to get it modified to reduce emissions. That is why the Clean Air Act requires preconstruction review.

Process design information: The first step in the analysis is the identification of every unit within the plant that will emit any regulated pollutant, if the plant *as a whole*, after controls, potentially will emit the pollutant in greater than *de minimis* amounts defined in the regulations.

As a reminder, the "regulated pollutants" for purposes of PSD include the six criteria pollutants, four hazardous pollutants (arsenic, beryllium, mercury, and vinyl chloride), and five pollutants regulated in various new source performance standards (fluorides, sulfuric acid mist, hydrogen sulfide, total reduced sulfur, and reduced sulfur compounds). Both controlled and fugitive emissions must be considered, for each of the components of the plant.

When the Environmental Protection Agency prepared its draft pollution control guidance document (PCGD) for indirect liquefaction plants, it identified about 25 large sources of air pollution emissions within a single facility. Some of these sources emitted more than just one regulated air pollutant. In addition, the PCGD identified dozens of other lesser sources of air emissions.

Once a piece of equipment has been identified as giving off a regulated pollutant, then the analysis moves on to consider types of controls which might be applied to it. A synfuels plant will contain many different kinds of *emissions units.* Some will be fairly standard pieces of industrial equipment such as the coal-fired boiler used to raise steam or generate electricity, coal piles, or the cooling towers. EPA may well have established NSPS standards for units like these when they appear as components of other industrial facilities. Those standards are not binding in a synfuels plant BACT review, but they do provide some guidance. Other emissions units that have not been built before in the configuration proposed for a synfuels plant will require new control technology.

Having identified emissions units which are sources of regulated pollutants, the next step in the BACT analysis is to identify "potentially sensitive concerns." This term is used to indicate specific air quality problems which might arise in the area of the proposed plant. If, for example, the plant is located in an agricultural area, the impact of emissions on crop productivity would be such a concern. Similarly, impacts on visibility would be important in a scenic area. The identification and, where possible, quantification of these concerns can be used later to help in the final decision balancing the costs and benefits of various alternatives.

Which brings us to the next, and perhaps most crucial, step in the process, the *selection of alternative control strategies.* For each unit, the applicant must identify different control strategies which might reduce the emissions of regulated pollutants from that unit. Normally, the applicant will identify a *base case* which consists of the control measures that would apply if there were no BACT requirement. In addition to the base case, the applicant must analyze at least some other plausible alternatives applying more stringent reductions.

The applicant must be prepared to demonstrate the technical feasibility of the base case or any alternative strategy it analyzes. This demonstration is generally done by making reference to the performance of the strategy on identical or similar processes. Control techniques that have not yet been demonstrated may also be examined. They are called "innovative control technologies," and the PSD regulations put certain restrictions on their use. Innovative control technologies are likely to be common where synfuel plants are concerned, because so many of the emissions control problems are new. But one should make sure that inexpensive and possibly insufficient control technology is not proposed as BACT simply because no other alternative has been demonstrated.

Impacts analysis. Once control alternatives have been identified, the applicant must analyze each to determine: how much the alternative will cost, (2) how much energy the alternative will use, (3) how much of each regulated pollutant the alternative will emit.

One critical function played by the impacts analysis is to study *incremental effects* of various options. A particular technology may be useful in removing 90% of projected emissions from a unit at a set cost. To get to 91% might require a second technology which is more expensive. But that second technology might have the ability to remove up to 99% of the emissions from the unit, with the additional 8% essentially free once the technology is installed. (This is the case, for example, with venturi scrubbers and electrostatic precipitators, used for removing particulates.) The impacts analysis must look at the incremental economic, energy, and environmental costs of going from 90% to 99%, not simply from 90% to 91%.

Each impact analysis must assess the effects of control options on the costs of the project and on any of the potentially significant concerns identified in the permit application. Again, two control options might allow almost the same amount of particulate matter to escape, but one might almost entirely eliminate those particles that produce haze and visual distortion, while the other might not. Since visibility is an important concern, this information should have an effect on the selection between alternatives. Alternative strategies may produce exactly the same pollutant concentrations, but one might not reduce emissions—this is the case when tall stacks rather than scrubbers are proposed to reduce SO_2 concentrations. And the analyses should also flag important environmental issues which do not just involve air quality. Certain control devices minimize air pollution, but dramatically increase water pollution or solid waste.

Choosing the best available control technology: The completion of the impacts analysis sets the stage for making the actual BACT determination. BACT is defined in the regulations as:

> an emissions limitation based on the maximum degree of reduction for each pollutant subject to regulation under the Act which would be emitted from any proposed major stationary source . . . which the reviewing authority, on a case-by-case basis, taking into account energy, environmental and economic impacts, and other costs, determines is achievable for such source . . . through application of production processes or available methods, systems and techniques . . .

The regulations further specify that BACT cannot be less stringent than any new source performance standard or hazardous emissions control which might apply to the source. None currently applies to synfuels plants.

It is clear from numerous court decisions interpreting the term BACT, that the decision maker must perform a balancing test. BACT is not really the best available technology; it is the best technology, taking costs into account.

The applicant will choose its proposed BACT and make it the basis of its air quality analysis. It must show, at a minimum, that, using its proposed BACT, it will not violate any NAAQS or the available increments. The reviewing authority need not accept the applicant's proposed BACT, but so long as NAAQS and increment consumption limitations are met, it has considerable leeway to approve any proposal. That leeway is greatest where noncriteria pollutants are involved, because of the lack of ambient air quality standards or increments to guide decision making.

When the BACT analysis is complete, it will indicate just which choices the synfuel developer wishes to make in selecting control options. But examination of the control options and the impacts analysis should give interested citizens some insights into the costs and benefits of more stringent and protective alternatives. You may be able to use the applicant's analysis to justify such alternatives or—if you find that costs for alternatives were improperly overstated or benefits underreported—attack the reasoning that led to a rejection of such alternatives.

The air quality analysis: Once the applicant has selected its preferred control technologies, it must show that the proposed plant will not cause a violation of the NAAQS for any criteria pollutant and that it will not exceed the PSD increment for particulates or sulfur dioxide. It must also show that there will be no adverse impact on any nearby Class I area.

To fulfill this requirement the applicant's air quality analysis must:

(1) determine the area within which air emissions from the plant will have an impact;
(2) ascertain the existing air quality within that impact area;
(3) demonstrate the effect that the additional emissions from the proposed plant will have on existing air quality.

The applicant must first define the area within which the plant's emissions of each regulated pollutant will have an impact above minimum levels established in the regulations. This is accomplished by the use of theoretical models that attempt to calculate the effects of local weather conditions and topography on the dispersion of emissions.

Once the impact area for each regulated pollutant has been specified, the applicant must determine existing ambient air concentrations. This is one of the most controversial aspects of the PSD program because the regulations provide that, as a general rule, the applicant must provide up to one year of continuous air quality monitoring data for each criteria pollutant emitted in significant amounts. As a practical matter, however, a full year's monitoring has rarely been required.

If on-site monitoring is required for any regulated pollutant, that effectively pushes the time at which an application can be made back by 14 months (a month to prepare, a year to monitor, a month to analyze). In addition, on-site monitoring is expensive. So it is in the applicant's interest to avoid the on-site monitoring requirement.

The reviewing agency may grant an exemption from monitoring requirements if the proposed emissions of a specific pollutant are very small. In addition, if existing monitoring data are representative and complete, new monitoring is not required. In many areas of the country, federal, state, and local agencies have been monitoring ambient air quality for some years and those data might meet the requirements for collecting air quality data. But accurate monitoring is highly dependent on the location and quality of monitoring stations. A receptor one valley away from a plant may not provide important information.

EPA requires that existing data meet the same requirements that would be applied to on-site monitoring by the applicant. Requirements for monitoring are found in EPA's "Ambient Monitoring Guideline for Prevention of Significant Deterioration" (EPA-450/4-80-012) and in regulations codified as Appendix B to 40 CFR 58. If an applicant has been given permission to use existing monitoring data for one or more criteria pollutant, check the administrative record to see that the reviewing agency has made a proper finding that the submitted data is sufficient, representative, and complete. If the validity of this data is a particularly important issue (and it will be wherever sources are located in irregular terrain), carefully examine whether the existing data were gathered in accordance with the requirements of the monitoring guideline and Appendix B. If you believe they have not, be sure to raise this issue during the public comment on the application.

The reviewing agency can, but need not, require that monitoring data be gathered on noncriteria pollutants. There is little indication that this has been done with any frequency. For several of the pollutants, EPA has yet to certify adequate monitoring procedures. The regulations assume that emissions inventories and modeling will allow adequate information on the air quality impacts of noncriteria pollutants in most instances.

EPA or the authorized state agency also has the right to order *post-construction monitoring*. This power allows EPA to guarantee that no NAAQS or increment violations occur and additionally to check on the accuracy of the modeling effort.

The emissions inventory: In the process of preparing the air quality analysis, the applicant must compile an emissions inventory. The inventory is very simple. It is a listing of important emissions sources within the impact area or an outside source which affects air quality inside the impact area.

The material listed in the inventory is almost always necessary to perform accurate modeling of the ambient air quality within the impact area. It is absolutely necessary to determine how much of the particulate or SO_2 increment has been consumed already and whether the new source will violate the increment.

Generally, information on how much the other sources on the inventory emit can be obtained from EPA and state permit files. Every large source of emissions has some emissions standards it must meet. These are generally called the *allowable emission.* Where modeling based on allowable emissions turns up results that the applicant would rather not find, the applicant may argue that actual emissions are really lower and seek to model based on alleged actual emissions. Keep a close eye on this. If there is no legal requirement imposed to keep the actual emissions from increasing, then the result could be dirtier air.

Air quality modeling: The applicant must then use a mathematical model to analyze the actual effects of the plant on air quality. The model must assess the interaction of weather, topography, plant design, and existing ambient levels.

All of the information is fed into an EPA-approved dispersion model. The modeling exercise is not automatic. In addition to choosing the proper model or models, it is generally necessary to choose long- and short-term meteorology, consider how to handle tall stacks (if a plant builds a stack higher than EPA-determined good engineering practices allow to spread the pollution more thinly over a wider area, it is modeled as if it built the stack at the correct height), and analyze the impact of fugitive emissions (for which there are no good models). Each of these decisions involves choices which can either understate or overstate the impact of the plant.

Strictly speaking, modeling is probably only required for criteria pollutants emitted in significant amounts. The source must demonstrate that it will not violate the NAAQS for any of these. The permitting agency has the authority to allow modeling or other analyses of the impacts of increased emissions of any other regulated pollutants the plant may emit. Some analysis of the impact of emissions by the source of each regulated pollutant is required, however, so neither EPA or a state can give the applicant permission to forget totally about discussing the potential impacts of emissions of noncriteria pollutants. This may be important, because the noncriteria pollutants may be the most worrisome emitted by the facility.

Increment consumption: Even without the PSD program, new sources would have to demonstrate that they would not cause violation of the ambient air quality standards, though perhaps not in so rigorous a form. What is unique about the PSD program is its limitation on the emissions of some pollutants at levels well below the health standards. Here is manifest the goal of the program to *maintain* air quality. It does this by means of increment limitation. The bare bones of the program:

In each attachment or unclassified area, a*baseline* is established showing existing concentrations of particulates and sulfur dioxide.

All land is put into one of three classes.

Increments are established for allowable new concentrations of particulates and sulfur dioxide in each class.

No new plant can be built in an area if its operation would result in ambient air concentrations higher than the sum of the baseline plus the applicable increment for either pollutant.

The applicant must demonstrate that the plant will not cause ambient air quality to exceed the available increment for particulates or sulfur dioxide. The applicant's analysis must include all new emissions since the baseline was established, not just the emissions from the proposed plant. The emissions include:

(1) emissions from major sources that commenced construction after January 6, 1975;
(2) emissions changes occurring after the baseline date at sources whose previous emissions were included in the baseline concentration (if they have decreased, that amounts to a credit);
(3) emissions changes due to SIP revisions that are approved after August 7, 1977.
(4) minor and area source growth occurring after the baseline date.

When all of these have been aggregated, they will be fed into the model. Because pollution output varies so much due to weather and climatic factors over time, the key issue is often whether short-term standards are violated more than once a year in any location.

The analysis and the emissions inventory must take into account sources outside the impact area which may have an effect on air quality inside the area (thereby consuming increment).

Additional impacts analyses: Besides the BACT analysis and the air quality analysis just discussed at length, the PSD program requires that the applicant prepare three additional impacts analyses.

The first is called the *growth analysis.* It requires the applicant to project any associated industrial, commercial, or residential growth that will occur in the area, to estimate the air pollution emissions generated by this growth, and to perform an air quality analysis which includes the impact of the associated growth.

The synfuels plant is only a small part of the total growth that will come to the area in which it is located. It will bring workers, who will seek homes, and businesses to supply them with goods. Industries providing raw material (like a coal mine) and equipment will probably locate in the vicinity. All of this associated growth brings with it some pollution. The growth analysis is designed to help the permitting agency and the public get a better feel for the total air pollution picture after the synfuels plant is up and running.

The source must also provide an analysis of the impacts of the additional air pollution, including that postulated in the growth analysis, on soils and vegetation. Concentrations of criteria pollutants below the NAAQS can have an adverse impact on some types of vegetation. Some of the noncriteria pollutants have even more marked effects on vegetation at very low emissions levels.

And finally, the applicant should provide an analysis of the impacts that pollution from the new source will have on visibility within Class I areas. This is called a *visibility impairments analysis.*

Impacts on Class I areas: If the permitting agency receives a PSD permit application from any source that could impact a Class I area (generally a source located wthin 60 miles of an area), it must immediately notify the Federal Land Manager (the secretary of the department managing the lands: for National Forest wilderness areas, it is the Secretary of Agriculture; for all other Class I areas, the Secretary of the Interior) and the federal official who directly manages the land, such as a park superintendent.

If the Federal Land Manager demonstrates that the emissions from the plant would impair air quality-related values (the most obvious is visibility), even though they would not violate the Class I increment, he may request that EPA or the state deny the permit. If, in the language of the regulations, EPA or the state "concurs with such demonstration," then it must deny the permit.

Within 30 days after the filling of a PSD application, EPA (or the reviewing state agency) must advise the applicant whether the application is complete. EPA must make a final decision within one year after the application is deemed complete (sources competing for the same pollution increment take priority based on the data upon which the completed application was filed).

Reviewing the application: As soon as the application is certified complete, EPA can begin to review it. Generally a technical review is made for each pollutant emitted in significant levels. The technical review is normally divided into two steps: a control technology review and an air quality review.

The control technology review has two phases. It must determine that a new source meets any applicable new source performance standards or hazardous air emissions standards which might apply. Currently there are no such standards for synfuel plants. It must also assess the adequacy of the applicant's proposed BACT measures.

The air quality review checks to see that the applicant has correctly demonstrated that (1) particulate and sulfur dioxide increments are not violated, (2) all NAAQS are complied with, and (3) air quality-related values, Class I areas, and nonattainment areas will not be significantly affected.

While the technical reviews are a very important step in the decision-making process, the regulations do not provide for any formal public involvement at this step. A phone call or a visit to the permitting agency may provide you with all the information you need, but it is unlikely that information about this stage will appear in the Federal Register.

The preliminary decision: The perliminary decision marks the first real public step in the process. Based on the result of its technical review, EPA will make a tentative decision to grant or deny the permit and, where it chooses to grant, a further decision on just which conditions to impose as a part of the permit. In reaching the preliminary decision, EPA must prepare a draft permit or issue a notice of intent to deny, it must document the finding with a Statement of Basis (SB) or a Fact Sheet (FS), and it must prepare an administrative record.

The draft permit will usually contain an itemization of the permit conditions which EPA intends to apply to the proposed plant. It becomes the target of the public hearings. Examine it carefully; it should indicate exactly what kind of pollution control measures the applicant will have to use. By implication, you will also see what more stringent measures EPA has failed to impose.

Along with the draft permit, EPA must issue a documentation of its findings. This can take two forms: a short and normally inadequate Statement of Basis, which can only be used for permits which are not the "subject of widespread issues," or which do not "raise major issues"; or a longer Fact Sheet for those that do. Insist that a Fact Sheet be prepared for any synfuel plant.

The Fact Sheet must include descriptions of the type and quantities of pollutants the plant will emit, the degree of increment consumption, a summary of the basis for permit conditions, and reasons why requested variances or alterantives were or were not chosen. It must also describe the procedures to be used for reaching a final decision: the beginning and ending of any comment period, procedures for requesting a public hearing, any other procedures by which the public can participate, and the name and phone number of an agency contact for additional information. The Fact Sheet must be mailed to anyone requesting it. As soon as it becomes obvious that a synfuel plant will seek a PSD permit, send a request to be on the mailing list to the appropriate permitting office.

The draft permit must be based on the *administrative record* that EPA has compiled in processing the permit application up to that point. To enable either the applicant or citizens to challenge or evaluate any of the decisions embodied in the draft permit, EPA must make the administrative record available for inspection and review (including copying) at some central location. The record must include all material submitted by the applicant and the files compiled by EPA during the review process.

Public participation: Once the draft permit is issued, the regulations provide for public notice, public comment, and the opportunity for a public hearing.

The *public comment* period must last for at least 30 days from the date of the public notice. During that period, any person may comment in writing. The permitting agency must consider all comments when making a final decision and must respond in some fashion.

Any citizen may request, in writing, a public hearing on the PSD permit, if one has not already been scheduled. A public hearing cannot be held on less than 30 days notice, so if the public hearing scheduled in response to a citizen request falls after the originally scheduled date for the close of public comment, the public comment period must be extended until at least that date.

The permit regulations are a little vague on whether EPA has to grant a request for a public hearing, but Section 165(a)(2) of the act is clear—a public hearing must be held, if even one person requests it.

The public hearing is just one form of public comment. Written submissions to the permitting authority serve many of the same functions. Whichever method you choose to use during the comment period, you must raise issues during the comment period, if you want to be able to appeal later adverse decisions by the permitting agency. Raising an issue involves more than a mere mention: the issue must be raised and arguments and factual grounds supporting the issue must be presented. It may be nearly impossible to raise adequately complex issues during a short comment period. If one has been proposed and you need more time to deal with it, you may request an extension of the comment period.

The public hearing: The public hearing available to present views on a draft PSD permit is closely akin to a public meeting; it is an extremely informal process. EPA will designate a Presiding Officer to run the meeting and individuals can present their views.

In general, there will be no opportunity to question the EPA officials responsible for the draft permit, no opportunity to question the applicant, no chance to cross-examine or question any other participants in the hearing. However, you can request such procedures, if you think they would help; the agency can grant or deny your request.

A tape recording or transcript is made of the hearing, and anything said at the meeting will become a part of the administrative record.

The PSD public hearing may prove a valuable opportunity to present your views and make the best case for them. It is not a particularly useful forum for unearthing new information (that should be done ahead of time).

Responding to comments: After the comment period closes, EPA must reconsider the draft PSD permit in light of data, information, and arguments raised during the comment period. As part of the administrative record for the final decision, EPA must describe and respond to "all significant comments" on the draft permit.

If material submitted during the public comment raises substantial new questions about a permit (for example a demonstration that the wrong model might have been used), the permitting authority has several options. It may prepare a new draft permit and begin the public comment period on it. Or it may reopen the comment period for the purpose of allowing others to comment on the particular new issue which was raised.

The final decision: After the close of the comment period and the finish of any analysis or response to issues raised during that period, the permitting authority must make a final decision on the permit application. If its decision is to issue a permit, it will prepare a final permit containing all the conditions that will be imposed.

Administrative appeal: Within 30 days of a final decision made by EPA or a delegated state, ay person who submitted comments on the preliminary determination may petition the EPA Administrator to review the final decision and permit. You must make an administrative appeal first if you intend later to challenge a PSD decision in the courts.

Citizen Suits

The Clean Air Act (Section 304) gives citizens the right to sue a person or company after providing 60 days notice, if: (1) an emission standard has been violated, or (2) an order of the EPA Administrator or a state regarding an emission standard has been violated. Citizens can sue a potential polluter who has commenced construction without a permit, or one who is in violation of any condition of a permit.

Citizens also have the right to sue the Administrator of EPA for failing to perform nondiscretionary duties, i.e., those specifically defined and required by law.

PSD permit challenges: Challenging the issuance of a PSD permit, while not impossible, is very complicated, time-consuming, and expensive. Most legal challenges are unsuccessful. For EPA-issued PSD permits, the relevant law is Section 5307(b) of the Clean Air Act and the Consolidated Permit Regulations, 40 CFR Section 124.19, 45 Federal Register 33491 (May 19, 1980). The Clean Air Act allows citizens to sue the Administrator of EPA in a United States Court of Appeals if a permit has been issued which does not comply with the PSD requirements (BACT, monitoring, modeling, etc.). The Consolidated Permit Regulations, which are incorporated in the PSD regulations, require citizens to raise all the problems with the permit during public hearings and comment periods. Except if there are major variations between the draft permit and permit, citizens are precluded from raising any issues not previously addressed during the comment period.

If the PSD permit is issued by a state, there are several legal options. Citizens could sue in state court or in federal district court under Section 304 of the Clean Air Act. There are legal questions concerning just who can be sued under what circumstances; the potential polluter, the state, or EPA.

Unlike the NPDES water pollution permit, the PSD permit becomes effective as soon as it is issued. Legal challenges to a PSD permit will not, themselves, hold up construction of the plant unless citizens are specifically granted an injunction to prevent construction.

THE CLEAN WATER ACT

The Clean Water Act (CWA) protects the nation's rivers, streams, harbors, and lakes. It is administered primarily by the federal Environmental Protection Agency, but states with adequate programs can assume primary responsibility for its implementation. There is, however, one exception. Under Section 404 of the act, disposal of dredged and fill material is regulated by the U.S. Army Corps of Engineers.

We will first review relevant provisions of the Clean Water Act and then describe the process through which plant sponsors obtain discharge permits.

I. Basic Goals: No Discharge/Fishable Swimmable

Unlike the Clean Air Act, the Clean Water Act does not focus on the definition of national health standards. Instead the act sets two basic goals:

1. That the discharge of pollutants into navigable waters be eliminated by 1985.
2. That, wherever possible, an interim goal of water quality which provides for the propagation of fish, shellfish, and wildlife; and provides for recreation in and on water be achieved by July 1, 1983.

These two goals, often referred to as the "no discharge" and the "fishable, swimmable" goals, reflect the clear intent to make protection of our water a first priority. The goals will not be met, but substantial progress has been made. The Clean Water Act deals with a range of sources of water pollution including:

a) direct point-source discharges (effluents discharged in a collected stream through a pipe or ditch or sewer, etc.), including effluents from industrial plants and municipal sewage treatment works (Section 402);
b) indirect discharge of industrial wastes via municipal sewage systems (Section 307);
c) non-point discharges including urban and agricultural runoff (Section 208);
d) the dumping of dredged and fill materials into wetlands and other waterways (Section 404);
e) discharge of heated waters (Section 316) ;
f) spills of oil and hazardous substances (Section 311);

II. Effluent Limits and Technical Standards

While several sections of the act will apply to synfuels plants, the most important is Section 402, which regulates industrial point source discharges.

Industrial effluents are to be regulated for levels of:

a) "conventional" pollutants such as bacteria, sediment acidity, and oxygen consuming capacity; and
b) toxic chemicals and hazardous substances like mercury, cyanide, benzene, phenol, and PCBs.

The principal criteria for setting limits on discharges of pollutants are:

a) application of the "best available technology" economically achievable (BAT) employed anywhere by industry for controlling the kinds of toxic and hazardous pollutants present in the waste stream, including changes in plant design and operation as well as "end of pipe" treatment;
b) utilization of "best conventional technology" (BCT) for control of conventional pollutants, roughly equivalent to treatment provided by the average modern municipal sewage treatment plant;

c) application of "best management practices" (BMP) to control the release of toxic and hazardous chemicals resulting from plant site runoff, leakages, and drainage from raw material and waste storage sites;

d) prevention of discharges that would result in the violation of state-set (and EPA-approved) standards for the quality of the receiving waters, even if steps beyond BAT or BCT are required.

According to the Clean Water Act, EPA defines what BAT and BCT discharge levels for various chemicals will be for each of several dozen categories of industry. Unfortunately, EPA is far behind schedule in issuing these effluent guidelines; and hence, binding guidelines for the synfuels industry are not likely to appear until the mid-1980s. (Non-binding guidelines may be released in 1982).

In the absence of EPA-issued industry-wide effluent limitations for the synfuels industry, BCT and BAT will have to be defined on a case-by-case basis for individual plants. Permit writers for these facilities may draw upon a variety of technical information such as (1) proposed or final effluent guidelines for similar industries such as coke making and petroleum refining; (2) so-called "development documents" which provide the background information used by EPA in deriving said effluent guidelines; and (3) the EPA "treatability manual" which describes methods whereby levels of various pollutants in waste streams can be reduced.

NPDES Permits

At the heart of the water pollution control regulatory system is the National Pollutant Discharge Elimination System (NPDES), established by Section 402. Every "point source" of water pollution must receive a Section 402 permit before it can discharge any pollutant into navigable waters.

NPDES permits are issued by EPA or by the state in which a facility is located if the state has qualified to take over the regulatory role (33 states now issue NPDES permits). The NPDES permit imposes precise and detailed pollution control requirements. It limits discharges of pollutants based on BAT effluent guidelines, new source performance standards, and applicable water quality standards for the receiving waters.

Standards of Performance

Under Section 306 of the Act, EPA is allowed to establish new source performance standards (NSPS) for newly constructed facilities; the act specifies 27 categories of sources for which such standards will be developed. Other sources can be added. There are no Section 306 standards for synthetic fuel plants, nor are there likely to be any. If they are developed, they would be applied through NPDES permits. In practice, the standards of performance for new facilities have been equivalent to the BAT requirements developed for existing facilities of the same class. A facility constructed pursuant to a standard of performance cannot be subject to a more stringent standard for 10 years after its permit is first issued.

Water Quality Standards

As noted above, one element of the regulatory scheme is provided by ambient water quality standards. Section 303 of the act requires states to adopt standards for the cleanliness of rivers, lakes and other bodies of water within their boundaries. If the states fail to do so, EPA may make such designations.

The water quality standards must protect public health and safety and must also be in line with the 1973 "fishable, swimmable" goal. They must also protect the quality necessary for the continuation of existing uses of the receiving waters and provide protection for downstream water quality.

In practice, states establish water quality standards by first creating a classification based on present or proposed uses for a body of water. Though the nomenclature varies from state to state, streams will be classified for uses such as drinking water, swimming, fish and wildlife propagation, irrigation, and industrial use. Degradation which would render a body of water unfit for existing uses is not allowed. For streams and lakes which are "outstanding national resources," even minor degradation is not allowed by current EPA regulations.

After a stream has been placed in a specific use classification, chemical, physical, and biological water quality criteria aimed at protecting the designated stream must be met. Where application of current effluent limitations alone is not sufficient to maintain the applicable water quality standard, more stringent NPDES permit requirements are set. This is generally done by setting maximum daily pollutant loads for the water body consistent with the applicable water quality standard and assigning a portion of the load to each of the discharges.

III. Section 404: The Dredge and Fill Program

The Army Corps of Engineers has had regulatory authority over the disposal of dredged and fill material into navigable waters since the passage of the Rivers and Harbors Act of 1899. Section 404 of the Clean Water Act continues that authority.

While this program may seem to be far removed from the operation of a synfuels project, it can have an impact on construction of the plant or of associated facilities such as pipelines. The program is more extensive than might be thought because the definition of navigable waters covers almost any body of water in the country plus wetlands.

When the Clean Water Act was passed, the Corps interpreted its jurisdiction to mean those waters which had traditionally been classified as "navigable," i.e., subject to waterborne commerce. As a result of a lawsuit, *Natural Resources Defense Council* v. *Callaway,* the Corps was ordered to expand its regulation beyond the traditional definition of navigability. Waters covered by the 404 program are now grouped into three classes. Phase I waters are traditional navigable waters and their adjacent wetlands; Phase II waters include primary tributaries of traditionally navigable waters and natural lakes greater than five acres in surface and their adjacent wetlands; and Phase III waters are all waters up to the headwaters where stream flow is less than five cubic feet per second. In short, if you either dig up or dump into any moist spot in the land, you are probably covered by the dredge and fill program.

The Secretary of the Army (who actually signs the permits in his role as head of the Corps of Engineers) can issue both general and specific permits. General permits can be issued for a class of activities which are similar in nature, cause only minimal adverse environmental effects when carried out separately, and will have only minimal cumulative effects on the environment. No general permit can last longer than five years.

Specific permits are tailored to individual projects. They must ensure that the activities involved in the dredge and fill operation do not cause water pollution or impair navigability. A state may administer both general and individual dredge and fill permit programs involving Phase II and III waters after approval of their program by the Administrator of EPA. A state must have essentially the same powers and laws it needs to administer the NPDES program to assume its limited authority over the 404 program.

Dredge and fill operations often accompany roadbuilding and construction, particularly near the coast, in wetlands, and in areas with many streams and lakes. Although temporary roads for moving mining equipment do not require 404 permits, other access routes do. Where such operations are part of a project that would have a significant impact on the environment, the Corps may have to prepare an environmental impact statement.

IV. NPDES Permit Procedure and Timetable

A. When must the permit be acquired?

For a new discharger, like a synfuels plant, a permit must be acquired before any pollutants can be released into a waterway. This must be a final permit, not a draft; and discharges cannot commence until all challenges to the final permit have been resolved, except in certain limited situations. (See discussions of hearings, appeals, and legal challenges below.)

B. When must the application be filed?

New dischargers must file their applications for a permit 180 days prior to commencing discharging.

C. What must be included in the application?

Among the most important pieces of information to be included in an NPDES application are:

—the name and exact location of the facility;
—the nature of the business to be engaged in at the facility, including what will be manufactured;
—the manufacturing processes to be employed;
—a line drawing showing the flow of water through the plant;
—the exact location, flow rates, flow frequency, and chemical contents of each discharge associated with the plant; and
—the wastewater treatment to be applied to each waste stream.

Existing facilities are required to submit the results of detailed chemical analyses of their waste streams, giving the concentrations and total daily discharges of some 160 chemicals listed on the application form. Applicants also are required to list and identify the source of any of 73 additional hazardous pollutants they know or have reason to believe they are discharging.

Since all synfuels plants will be coming in as new dischargers, there will be no actual effluent streams to analyze. Applications will, however, be required to provide estimates of concentrations and total discharges of any of the 160 key conventional and toxic pollutants expected to be discharged, and to give the name and source of any of the 73 hazardous pollutants likely to be present in the wastewaters.

If an applicant fails to submit a complete application, processing will be held up until the needed information is provided. (Deliberate falsification of information on an NPDES application is a criminal offense subject to fines of up to $10,000.)

The most serious possibility of an omission of important information on the application arises if a facility would discharge a highly toxic chemical that does not appear on any of the lists included with the application form. The public might not learn of this danger associated with the facility's operation until the chemical's effects on human health and/or the environment had been directly demonstrated.

NOTE: The above information applies to applications submitted directly to EPA, on its Forms 1 and 2C. In cases where the state is running the permit program, there might be some slight variation in the requirements.

D. Agency Review Process

Once EPA or a state agency has what it believes to be a complete application, the process of analyzing the information provided and developing specific effluent limits for various pollutants can begin. Some agencies have a policy of discussing permit issues with the applicant and interested members of the public, while others tend to operate privately until the draft permit is issued. If possible, it is probably wise to get involved in the permitting process before issuance of the draft permit, as once an agency has released a draft permit, it usually feels it has a stake in defending it. (Some would argue, however, that one can be co-opted by getting involved in private negotiations with the agency and permittee.)

In considering whether to issue or deny a permit, and what conditions to place on the permit if issued, the permitting agency will consider a number of factors. These include the previously mentioned criteria for application of BCT, BAT, and BMP. Based on these considerations, the application will be denied or a draft permit will be issued.

1) Permit Denial

A permit cannot be issued if:

a) imposition of conditions still cannot ensure compliance with relevant state water quality standards;
b) the Army Corps of Engineers objects to the permit because it would impair anchorage or navigation;
c) the EPA Regional Administrator has objected to issuance of a permit by a state; or
d) the state in which a proposed discharge is to be located refuses to certify that the discharge will comply with the Clean Water Act and related state statutes.

2) Issuance of a Draft Permit

If the permitting agency determines that a permit should be issued, a draft permit is prepared. In addition to providing authority for an applicant to discharge into a waterway, a permit should include a number of conditions including:

a) *Effluent limitations for specific conventional pollutants, toxic pollutants, and hazardous substances.* These are based on levels achievable by application of BCT, BAT, and BMP. Even lower pollutant levels must be required if the state's water quality standards will not be met by application of the above technology-based standards. Limits are expressed in terms such as pounds of pollutant released per ton of raw material used or product generated, milligrams of pollutant per liter of discharge water, or grams released per discharge (if not continuous). Limits are based on daily maximums and monthly averages.

Under some circumstances, levels of certain pollutants may be used as indicators of the application of adequate treatment for controlling a number of other chemicals, rather than setting individual limits on each of the involved pollutants. The purpose of utilizing such "indicator" pollutants is to reduce the costs to industry of monitoring discharges.

When limits on specific chemicals are not feasible or adequate, limits may be set on the basis of the overall toxicity of the discharge, as determined by biological testing.

b) *Monitoring*. Permittees are required to monitor regularly the levels of discharge of all pollutants for which a specific effluent limit has been written into the permit. Monitoring must be performed according to procedures outlined by EPA. Testing of the overall toxicity of a discharge may be used as a supplement to chemical monitoring in some instances.

c) *Reporting*. In addition to providing regular discharge monitoring reports (DMR), permittees are required to report to the responsible agency in a number of special circumstances including:

1. anticipated or unanticipated diversion of waste streams from treatment facilities;
2. malfunctions of treatment systems;
3. discharges of pollutants exceeding any effluent limitation in the permit;
4. planned changes in the plant or its activities that may result in noncompliance with permit requirements or cause significant increases in the levels of toxic chemicals not regulated in the permit.

d) *Term of the Permit*. Permits for synfuels plants are good for five years, unless conditions justifying permit modification or termination arise. (See Section I(3).)

Public Notice of Draft Permit Issuance or Permit Denial

State and federal permitting agencies are required to notify the public of the issuance of a draft permit in ways including:

a) publishing a notice in a newspaper within the area affected by the proposed facility; and
b) mailing a notice to the permit applicant, federal and state fish and wildlife and historic preservation agencies, state water resources planning agencies, the U.S. Army Corps of Engineers, and persons who have asked to be placed on a mailing list for notices. (Adjoining states with waters whose quality may be affected must receive a notice.)

The notice of permit issuance must include the name and address of the applicant, a brief description of the facility in question, the location of each proposed discharge point, the amounts of various pollutants to be discharged, a summary of the legal basis for the effluent limitations and other permit conditions, and an explanation of how the effluent limits were calculated.

The notice is also required to include a description of the procedures to be followed in reaching a final decision on the permit, including:

a) time allowed for submission of written comments on the draft, which can be no less than 30 days;
b) procedures for requesting a public hearing, or date and location of such a hearing, if already planned.

The notice must also include information on where to obtain additional information about the permit and the permitting process, including where and how to obtain copies of the permit application, draft permit, and other relevant documents.

E. Written Comments/Public Hearing

The permitting agency is required to give the public the opportunity to comment in writing on draft permits. Those wishing to contest a permit or its conditions must make their views known at this stage of the process. Issues not raised in written comments or testimony at a hearing cannot be raised in legal challenges or other appeals of a final permit. (See discussion of appeals to the Administrator, evidentiary hearings, and court challenges below.)

A public hearing is required on any draft permit that is the object of a significant degree of public interest—a virtual certainty in the case of a synfuels plant. The public must be given at least 30 days notice of a public hearing. The format of these proceedings is that of typical hearings before local, state, or federal legislative bodies, with people presenting oral and written testimony to representative(s) of the decision-making body.

If any substantial new questions are raised about the permit in writing or at a public hearing, the agency may extend the comment period or even issue a new draft permit and start the public input process over again.

F. Issuance of a Final Permit

After the comment process is completed, the permitting agency issues a final permit, reissues a new draft, or denies the permit. The final permit incorporates any changes the agency chooses to make based on input received during the comment period. (If EPA is the permitting agency, terms specified as conditions for obtaining state certification would have to be included.)

NOTE: State permit writers are required to submit a final permit they propose to issue to the Regional Director of the EPA for review. The EPA has 90 days to consider the permit, and must notify the state of its approval or disapproval. In the latter case, reasons for disapproval and changes needed to obtain approval must be spelled out. EPA must hold a public hearing to receive comments on its position if the state requests, and should also hold one if significant public interest is expressed. If the state refuses to alter the permit as requested, EPA can either deny the permit or issue it according to terms it sets.

Agencies are required to provide *notice of the issuance of a final permit.* In the notice, permit writers must specify which provisions of the draft permit have been changed and why, and must also describe and respond to all significant comments on the draft. Notice must be provided to the applicant and everyone who submitted comments or requested to be notified.

The *effective date* of an EPA-issued permit is 30 days after the notice of permit issuance, unless the permit is contested. (See below.) The permit is effective immediately upon notice if no comments objecting to the draft have been received.

H. Appeals and Legal Challenges

For EPA-Issued Permits interested parties are provided the opportunity to appeal a final permit.*

Quasi-Judicial Hearings. These proceedings, referred to by EPA as "evidentiary hearings," are much more formal than the public hearings, very much resembling courtroom proceedings with expert witnesses and the like. Participation in an evidentiary hearing requires much more time and money than testifying at a public hearing.

*State NPDES programs are not required by federal regulations to incorporate procedures for appeals of final permits. (The lack of such requirements has been challenged by environmental organizations.) State programs, therefore, vary in the degree to which they provide opportunities for challenges of permits beyond the public hearing stage. Some have procedures parallel to the EPA appeals process described here, while others do not allow citizen groups to appeal to the courts.

If someone wishes to challenge a final permit, they must file a request for an evidentiary hearing within 30 days following the service of notice of EPA's final permit decision. The request must be very detailed, including papers which state the legal and factual issues to be raised, with supporting documentation. No issues that were not raised during the comment period on the draft can be used, unless new information has become available or new issues have been created by the changes from the draft to final permit.

The EPA Regional Administrator must decide whether to grant the request within 30 days of the end of the period allowed for requesting a hearing. A request can be approved in full or in part, with some issues being allowed and others not. Reasons for denying a request must be stated. Notice of a decision on a hearing request must be given to everyone who provided written comments or testimony on the draft permit, the permittee, and the party requesting the hearing.

A denial in whole or in part of a request for an evidentiary hearing can be appealed to the Administrator of the EPA in Washington, D.C. within 30 days. Other parties to the proceeding can file petitions supporting or opposing the appeal. Within "a reasonable time" the Administrator must issue an order granting or denying the appeal. The appeal can be approved in whole or in part.

If an evidentiary hearing is granted, interested persons have 15 days to submit a request to be admitted as a party.

The proceedings are run by an administrative law judge appointed by EPA, and involve the filing of formal documents, cross-examination of witnesses, and other quasi-judicial procedures. Evidentiary hearings can take months and months, although much of the time is consumed in waiting for the filing of briefs, testimony, motions, and other papers by the parties to the case. In fact, the entire proceeding may take place in writing, with no actual "hearing" ever occurring. Although not required, representation by a lawyer in these proceedings is extremely helpful.

After reviewing the arguments and information presented in the proceedings, the administrative law judge issues a decision. If no appeal is filed within 30 days, the decision is final.

Appeal to the EPA Administrator. Appeals of the decisions rendered in an evidentiary hearing must go to the Administrator of EPA. The process is identical to that for appeals of a denial of a request for an evidentiary hearing (see above). The Administrator can affirm the decision of the administrative law judge, alter the permit and issue it as final, or send the case back to the administrative law judge.

Appeal to the Courts. The decision of the EPA Administrator on an appeal of an evidentiary hearing, denial, or decision resulting from an evidentiary hearing is appealable within 90 days to the federal courts. The courts can review only the record established during the evidentiary hearing process. (Note: one can appeal to the courts only after having exhausted other avenues of administrative appeals, i.e., going all the way to the Administrator of EPA.)

The permit is effective only upon the completion of all administrative appeals. The effectiveness of the permit can be stayed by court order.

I. Post-Permit Procedures

1) Monitoring and Reporting—Permittees' responsibilities for monitoring discharges and reporting to the government are spelled out in their permits. (See Sec. (D) (2) above.) Dischargers are given responsibility for measuring levels of pollutants in their effluent, but EPA and state agencies have the right to enter a plant to sample discharges, check monitoring equipment, and examine discharge data. (All effluent data obtained pursuant to an NPDES permit must be made available to members of the public.) Deliberate falsification of data on monitoring forms or other reports is a criminal offense subject to fines of up to $10,000, and possible imprisonment.

2) Enforcement—Violation of the terms of an NPDES permit is a civil offense under federal and state law, punishable with fines up to $10,000 per day of violation. Willful violation of permit conditions is a criminal offense subject to fines of up to $25,000 per day or imprisonment up to two years.

EPA has the authority to take direct enforcement action against a violator of a state-issued permit if the state fails to pursue enforcement.

Citizens (if they have an interest which may be adversely affected) can sue permit holders directly for permit violations if the state and/or EPA fail to take the necessary enforcement actions. The act also guarantees the right of citizens to sue EPA for failure to perform a non-discretionary duty, but no such right is provided with regard to state agencies, leaving this matter to state law.

Courts are authorized to award to litigants the cost of litigation (including attorney and expert witness fees) in association with suits brought with regard to permit violations. (Note: These costs are not authorized in appeals of permit decisions, described above.)

3) Modification of Permits—In general, compliance with an issued NPDES permit protects a discharger from more stringent pollution control requirements resulting from lawsuits or new regulations. However, permit requirements can be changed before a permit needs to be renewed under certain circumstances, including: emergence of important information not available at the time of permit issuance; substantial changes in the permitted facility or activities thereof; major increases in discharges of chemicals not limited in the permit; failure to comply with any condition in the permit; misrepresentation of facts in the permit application or elsewhere; or determination that the permitted discharge levels actively endanger human health or the environment.

Significant permit modifications must be subjected to the same opportunities for public input and appeal as an original permit issuance.

V. Public Participation in NPDES Process

In general, the NPDES process provides substantial opportunities for interested members of the public to provide input to the permit-writers, although the actual degree of participation allowed varies from state to state. It is, of course, hard to predict the chances of actually influencing the outcome of the permitting process.

Considering the political and financial power behind most synfuels plants, it is probably unlikely that an NPDES permit would actually be denied, but well organized and informed citizens could well be able to use the NPDES process to force additional necessary effluent controls on the facility, delay its construction, or at least bring to light information that would raise public concerns about the plant.

A. Gathering Information

The first step in any successful public participation effort is the collection of relevant information. As noted before, federal law and regulations guarantee public access to NPDES permit applications, draft and final permits, and effluent data on applications or monitoring reports. Apparently some states have refused to release some of this information in spite of the federal regulations. Since the state is supposed to send most of this information to EPA, this source should be checked as well. Of course, legal action could be considered if all other means of collecting this information fail.

Some synfuels plants will have overall environmental impact statements prepared, which should be examined for additional information. (For instance, some chemicals released by synfuels plants are not included on the lists of pollutants that must be reported on NPDES applications.)

Other EPA and state documents will be useful in analyzing an NPDES application, including EPA development document and effluent guidelines for related industries, and state water quality standards.

B. Commenting on Permits in Writing or at Hearings

1) Analyzing Permit Applications and Draft Permits

a) Permit Applications—Although federal and state regulatory bodies should ensure that applications are complete, citizens should double-check just in case.

—Watch for failure to include chemical test results when required.
—Make sure all outfalls and their contents are listed.
—Look for suspiciously low projections of levels of discharges of particular pollutants.
—Check for chemicals not listed on the application form that have been identified as likely effluents in other documents.

If there are any inadequacies in the permit application, press the regulatory agency to hold up processing of the permit and ask the applicant for the needed information.

b) Draft Permits—Although it might seem that an average citizen would not be able to analyze the contents of a draft NPDES permit, application of common sense and a little effort can produce results. It would, of course, help to be assisted by someone with knowledge of wastewater treatment, aquatic ecology, or toxicology.

Among the things to check in a draft permit are:

—Do all effluents listed in the permit application have limits? If not, why? Make sure the limits are not just in terms of pollutant concentration because pollutants could otherwise be diluted with clean water; meaningful discharge limits should be in terms of total amount of pollution discharged. Are there limits on maximum daily and average monthly discharges? Are limits placed on all outfalls or related internal streams?

—Do the limits on individual pollutants reflect limits proposed in EPA guidance documents for this kind of facility or limits in proposed or final effluent guidelines for similar industries?

—Do the effluent guidelines reflect application of the Best Available Technology for limiting pollutants, i.e., the most effective systems that have been applied by industry anywhere (including foreign countries) to similar waste streams?

—Will the proposed limits assure the meeting of all relevant state water quality standards, in nearby states affected by the discharges as well as the state in which the plant is located?

—Does the permit prohibit the bypassing of wastewater treatment facilities, except to prevent loss of life or extensive property damage? Are backup sources of power for treatment facilities required?

—Is it made clear that the permittee would be required to reduce or halt operations if necessary to prevent violation of permit conditions, even if this involved significant costs?

—Are there requirements for Best Management Practices to control leaks, spills, and runoff from storage piles?

—Are there adequate monitoring requirements? (For a synfuels plant it would not be unreasonable to require a full-spectrum chemical analysis of all relevant outfalls on a monthly basis. Frequent biological testing of the effluents for overall toxicity should also be required as a backup to the chemical tests.) If the plant discharges effluents on an intermittent basis, do the monitoring requirements clearly require testing to take place during representative discharges?

In addition to preparing your own comments on a draft permit, be sure that other relevant interests are doing likewise. Pay particular attention to federal and state fish and wildlife agencies, water resource planning agencies, and historic preservation offices. Elected officials are also important. Environmental protection agencies from nearby states affected by the project are also key. Do not forget to talk to state officials responsible for certifying EPA-issued permits, or persons in the EPA regional office responsible for reviewing state-issued permits.

In talking with the people mentioned above, do not assume that they know everything you do. Most agencies are understaffed and can not give adequate attention to most tasks. Also try to learn as much as you can from them. You may be able to collect scattered bits of information that together point to serious issues.

What about a hearing? In theory, information presented in writing would have just as much impact on the permitting agency as that presented at a public hearing. In reality, however, one must factor in the political pressure that can be put on an agency by having large numbers of people show up at a hearing to express their concerns. In addition, information presented at a hearing is more likely to get picked up by the press.

REMEMBER: If you want to challenge an NPDES permit, you must raise the issues that concern you at this stage of the process. Matters not introduced at this stage cannot be brought up in appeals of final permits unless it can be shown that it would have been virtually impossible to do so during the public comment/hearing phase.

c) Appealing a final permit—The evidentiary hearings that are employed to deal with appeals of EPA-issued final permits are much more lengthy and complicated than public hearings. Costs for duplication of briefs and testimony can be high, and legal representation is highly desirable, adding another potentially large cost factor. (In the past, EPA's regional offices have often responded to requests for evidentiary hearings from dischargers by modifying the permit in a way favorable to the discharger, in exchange for dropping of the hearing request. It is possible that citizens groups could extract similar concessions, thereby obtaining positive results (or at least discourage weakening concessions by the agency by becoming "parties" who must be consulted in any such settlement) while avoiding the substantial expenses of a full-blown hearing.)

What about further appeals? The outcome of an evidentiary hearing on an EPA-issued permit can be appealed to the EPA Administrator, and from there to the federal courts. (As explained previously, procedures for appealing state-issued permits are not necessarily the same as for those issued by EPA.

d) Post-Permit Monitoring—Because synfuels production involves the application of relatively untested technology, monitoring the *actual* effluents from a plant that has received an NPDES permit is extremely important, particularly during the first year or two of its existence. Dischargers are required to send Discharge Monitoring Reports (DMR) to the agency that has issued the permit on a regular basis specified in the permit. Permittees must give agencies reports on unanticipated bypasses, upsets, and violations of maximum daily discharge limits within 24 hours of occurrence. Significant increases in discharges of toxic pollutants not controlled in the permit must also be reported. Agencies also issue quarterly reports on incidents of noncompliance with permit conditions by dischargers. Concerned citizens should examine these reports to see if permit conditions are being met or if new conditions need to be imposed. (See Section I for further discussion of monitoring, enforcement, and permit modification.)

RESOURCE CONSERVATION AND RECOVERY ACT

Every day a synfuel plant will produce thousands of tons of solid wastes. One of the newest of the major environmental laws—the Resource Conservation and Recovery Act—provides the framework for federal regulation of disposal of solid and liquid wastes which are normally buried or stored.

The Regulatory Framework

RCRA (pronounced "rek-ra") was designed to close the circle of environmental regulation by controlling the handling of solid wastes. The centerpiece of RCRA regulation is found in Part C of the statute, which provides for a comprehensive "cradle to grave" coverage for *hazardous wastes.* It is important to recognize that Part C of RCRA applies standards only to hazardous wastes as defined. Other solid wastes, regardless of their potential enviromental impact, are *not* regulated by Part C.

Where a waste is identified as hazardous, RCRA creates a series of obligations for the generator of the waste, for anyone who transports the waste, and for the person who stores or otherwise disposes of the waste. At the heart of this set of obligations is a *manifest* system. A generator of solid waste must prepare a manifest for all shipments of hazardous wastes that are sent to offsite treatment, storage, or disposal facilities. The manifest serves as a tracking device, to provide relevant information to persons handling the waste, and to provide a proper record.

The RCRA regulations encourage generators of hazardous wastes to ship their wastes—under manifest—to treatment or disposal facilities as soon as possible. If a generator accumulates wastes for more than 90 days prior to shipment, it is treated as someone storing wastes and is required to obtain a facility permit and operate under regulations governing the operation of hazardous waste disposal facilities.

If wastes are shipped offsite, the manifest system is designed to enable the generator and EPA to keep track of the wastes and assure that they are disposed of at a permitted facility. If a copy of the manifest is not returned to the generator within 45 days of shipment, an exception report must be filed with EPA. Some states require a copy of all manifests to be submitted to the relevant state agency and have shortened the exception reporting period.

Treatment and storage facilities are subject to a series of requirements. Among these are administrative requirements: the storage facility must maintain adequate records, must comply with the manifest system (by reporting final disposal of the waste) and must meet certain standards for training personnel and demonstrating financial responsibility for continuous care of the site. There are also specific performance standards for the operation of the facility which are designed to insure that the stored or treated hazardous wastes do not reenter the environment as pollutants or poisons. No treatment, storage, or disposal facility may operate without a permit, granted by EPA or a state which has assumed permitting responsibility pursuant to RCRA.

The storage or treatment facility permit defines more specifically what can and cannot be done in disposing of wastes, which hazardous wastes a facility may accept, and what types of treatment are allowed.

If a synfuel plant were to generate quantities of hazardous wastes, it would either have to ship them to a storage facility or—if they were stored at the site—obtain certification and permits as a storage or treatment facility.

The cradle-to-grave treatment of Part C of RCRA only applies to wastes defined as hazardous. According to the interim final regulations published on May 19, 1980, a waste may be defined as hazardous in either of two ways. It may meet standards set forth in those regulations for ignitability, corrosiveness, reactivity, or EP (extraction procedure) toxicity; or it may be *listed* as a hazardous waste by EPA.

Hazardous wastes are listed individually and by industrial process. If a waste is hazardous, then any mixture of that waste and other wastes is also treated as hazardous. EPA has also included a long list of hazardous constituents which are candidates for inclusion on the list of hazardous wastes in the future.

The application of Part C of RCRA to synfuel facilities is highly problematic. The four criteria for inclusion deal primarily with short-term acute effects. Some synfuel solid wastes that have been tested seem to pass the tests with little difficulty. EPA has not established criteria for carcinogenicity, mutagenicity, teratogenicity, bioaccumulation potential, or phytotoxicity. Rather, it seeks to cover the more important wastes having these characteristics by listing. Thus, unlisted compounds with these characteristics will not be considered hazardous if they are not produced by a listed process. Unlisted compounds which happen to be ignitable, corrosive, reactive, or exhibit EP toxicity (which is the ability to leach out concentrations of materials subject to primary safe drinking water standards) will be treated as hazardous because generators are required to test for those characteristics and, on finding them, notify EPA. Applying the primary drinking water regulations as the test for EP toxicity has major shortcomings, since there are only 17 drinking water standards and almost no toxic organic chemicals are covered. EPA justified its failure to include *etiological* characteristics such as carcinogenicity "because of the lack of suitably uncomplicated test protocols, the difficulty of establishing numerical threshold levels, and the failure of the available test protocols to fully incorporate all the multiple factors bearing on the hazards presented by such characteristics." Many hazardous end products of synfuels production are on the hazardous constituents list and could be covered by RCRA, although they are presently outside the system.

In the short run, there is a more definite barrier to the inclusion of most synfuel wastes within the purview of Part C— even if they happen to meet the standards for ignitability, corosivity, reactivity, or toxicity. In 1980, Congress passed the Solid Waste Disposal Act Amendments (P.L. 96-482). A little noticed provision of the act prohibited EPA from using RCRA to regulate:

> (i) Fly ash waste, bottom ash waste, slag waste, and flue gas emission control waste generated primarily from the combustion of coal or other fossil fuels.

(ii) Solid waste from the extraction, beneficiation and processing of ores and minerals . . .

until a study is completed on the environmental effects of the disposal of such solid wastes. Even after the study is completed, RCRA will apply only if the Administrator of EPA decides to promulgate new regulations to include these substances under RCRA.

Wastes That Are Not Hazardous

Another section of RCRA may provide additional regulations for solid waste disposal practices. This is Part D of RCRA. Section 4005 of the act requires states, as a part of their approved plan for implementing RCRA, to phase out "open dumping" of solid wastes. An open dump is one that neither meets the criteria for storing hazardous wastes established under Part C nor the criteria for being a *sanitary landfill* (defined in Section 4004). The practical effect of this provision is to establish some minimal standards for the disposal of nonhazardous wastes. (If states do not take the carrot of adopting plans to phase out open dumping, they face the stick of having it banned immediately.)

State Programs

As with the other major environmental laws, RCRA provides a method by which states may assume primary responsibility for administration of the hazardous waste disposal program.

State Laws

States have authority to and some states have established strong independent programs to regulate solid wastes. Unless RCRA-defined hazardous wastes are a significant portion of synfuel-generated solid wastes, or individual states enlarge the definition of hazardous waste, these laws will be the primary governor of solid waste disposal practices.

TOXIC SUBSTANCES CONTROL ACT

The Toxic Substances Control Act (TSCA) is both a product control law—like the Food and Drug Act—and a pollution control law. It regulates the manufacture, processing, distribution, use, and disposal of unreasonably hazardous chemical substances and requires screening of new chemicals to identify those which "may present an unreasonable risk of injury to health and the environment." Under TSCA (pronounced "toss-ka"), EPA may also require that existing or new chemicals be tested for toxicity. Thus, where synfuel plants are concerned, TSCA can regulate the synfuel plant and its operation as well as the use of the synfuel product—the oil, gasoline, petrochemical feedstock, or other substance produced by the facility.

TSCA requires the manufacturers of "new chemical substances" to present EPA with available data on the potential health and environmental effects of that new substance before it is manufactured and distributed. It also allows EPA to regulate (including prohibition) the manufacture or use of either a new or existing chemical substance where it determines that such use presents an unreasonable risk of injury to health and the environment.

Under the authority of TSCA, EPA has the power to prohibit the manufacture of certain synfuel products, if it finds them to be dangerous. It also has the power to invoke lesser sanctions, such as limiting certain uses, providing notification of risks, or following certain quality control procedures.

Premanufacture Notification

Synfuel products are particularly likely to be subject to TSCA's premanufacture notification (PMN) requirements. Under Section 5(a) of the act, companies must notify EPA at least 90 days before they begin to manufacture or import a "new chemical substance" for commercial purposes.

New chemical substances are those which are not already included on EPA's Chemical Substance Inventory—a massive compendium of more than 45,000 known chemical substances that were already in use at the time TSCA was passed. Chemicals on the inventory are grouped into two classes. Class I substances are those which can be identified through their molecular structure, such as benzene, methane, or vinyl chloride. Class II substances are chemicals with varying composition, often known as complex reaction products. Conventional petroleum products, as well as most synfuel products, fall into this class. Once a chemical passes through a premanufacture review, it is added to the inventory and future manufacturers of that chemical do not have to file PMNs.

EPA has yet to decide how specific synfuel categories must be. Are products from different coal liquefaction processes (for example, gasoline made from H-Coal and EDS) similar enough so that listing of one removes the need to make the other undergo a PMN? Should different feedstocks be taken into account, such as different coal seams which are known to have relatively different chemical composition? EPA has decided that coal-based methane and methanol are existing chemicals which do not require a PMN.

Where a PMN is required, because the synfuel product is a new chemical substance, Section 5(d) specifies the information that must be reported. It includes "reasonably ascertainable" information on chemical identity, anticipated product volume, categories of use, byproducts resulting from use, manufacture, processing and disposal, estimates of workplace exposure and risk, and information on the manner or methods of disposal. The manufacturer is also required to turn over to the EPA any studies it may have conducted or is aware of on environmental or health effects of the chemical, but Section 5 does not require that the manufacture develop such information. (Under Section 4 of the Act, EPA can order specific testing, but this requirement does not automatically apply to each PMN; it is done by rulemaking.)

After the PMN is submitted, EPA has 90 days to act. It can seek an additional 90-day extension for good cause. If it does nothing, manufacture can commence and the substance becomes part of the inventory. Once it is on the inventory, it is no longer a new chemical substance.

Section 5(e) allows EPA to delay the manufacture of a new chemical substance because the data submitted in the PMN do not allow for a "reasoned evaluation" of its health and environmental effects, *and*

> (1) the chemical may present an unreasonable risk to human health or the environment, or
>
> (2) significant human or environmental exposure can reasonably be expected.

Invocation of Section 5(e) essentially can put a synfuel project on hold while adequate environmental and health information is developed. It clearly would give a project developer an incentive to do so quickly. But no amount of pressure can create long-term exposure data where there has not been any long-term exposure; by definition, this will always be true for new chemicals.

Regulating New Chemical Substances

At some point, even if Section 5(e) is used, EPA will have to make an assessment of the health and environmental impacts of a new chemical substance. If the agency does not act under Section 5 authorities, the chemical can go into manufacture and distribution and ceases to be subject to the authority of Section 5.

If EPA finds that a new chemical will present an unreasonable risk to human health or the environment, it can be regulated under Section 5(f). Section 5(f) allows EPA to use a variety of options from requiring that the substance be labelled to banning its manufacture. Section 5(f) must be invoked before the PMN review period (including any extension created by use of Section 5(e)) expires. The options available are identical to those provided by Section 6, which applies to regulation of existing chemicals.

Regulating Existing Chemical Substances

TSCA also allows regulation of chemicals that are no longer new. Section 6(a) allows EPA to apply regulation to *any* chemical which it finds poses an unreasonable risk to human health or the environment. In addition, Section 6(c) requires the Administrator of EPA to proceed under other laws administered by EPA to reduce such a risk if the Administrator first determines that the risk would be better mitigated under other laws, unless a discretionary finding is made that it would be in the public interest to use TSCA.

Section 6(a), as did Section 5(f) discussed above, presents a wide range of protective options which can be imposed by EPA, singly or in combination. Among the most important are:

- —Prohibiting or limiting the manufacture, processing, or distribution (a) in total amount, (b) in relation to a specific use, or (c) in a specific use above a concentration level determined by EPA.
- —Requiring that any article containing a substance or mixture by accompanied by a clear warning of its contents and carry instructions with regard to its use, distribution, or disposal.

—Requiring the manufacturer or processor of such substances to retain records of the processes used to manufacture them and to monitor and conduct any tests to assure compliance.
—Requiring the manufacturer to give notice of risks associated with the chemical to distributors and any other persons in possession of the chemical, including public notice through advertisement.

EPA also has the authority to order a manufacturer to change or alter the process used to make a chemical if there is evidence that this might lessen a risk associated with the product. EPA officials have suggested that, in light of this power, synfuel manufacturers might well undertake pre-production tests of chemical process changes such as substitution catalysts and combustion parameters—focusing on two process elements which often appear to influence the potential for toxic behavior in the end synfuel product.

To date, EPA has done little under TSCA with respect to existing chemical products from well-understood industries. Accordingly, it is difficult to predict whether TSCA will have an impact on the synfuels industry. But because some synfuel products, particularly substitutes for automobile gasoline, could have widespread distribution and concomitant human exposure, the potential for regulatory activity remains.

THE ENDANGERED SPECIES ACT

The Endangered Species Act authorizes the Secretary of the Interior, in consultation with the Secretary of Commerce, to promulgate regulations identifying *endangered species* and *threatened species.* An endangered species is:

> any species which is in danger of extinction throughout all or a significant portion of its range other than a species of the Class Insecta determined by the Secretary to constitute a pest whose protection under the provisions of this chapter would present an overwhelming risk to man.

A threatened species is:

> any species which is likely to become an endangered species within the foreseeable future throughout all or a significant portion of its range.

At the end of 1981, there were 681 (236 in the U.S.) endangered species and 75 (52 in the U.S.) threatened species listed by the Department of the Interior. The Fish and Wildlife Service has the responsibility for identifying animals or plants which might qualify for endangered species designation.

Prohibited Acts.

Section 9 of the ESA sets out specific prohibitions. Most concern deliberate killing or removal of the species: bans on hunting, trapping, killing, importing, exporting, or selling an endangered or, in most cases, threatened species. However, Section 9 also prohibits "harming" a listed species. "Harm" is defined by regulation as "an act which actually kills or injures wildlife. Such act may include significant habitat modification or degradation where it actually kills or injures by significantly impairing essential behavioral patterns, including breeding, feeding, or sheltering." [46 Fed. Reg. 54748-5.]

This definition has not been tested in court, so it is unclear whether it would apply to, for example, cutting down a bird's nesting tree during the non-nesting season. However, it is potentially a strong protection of habitat for *animals* (plant species are not protected from "taking," the category which includes "harm"). This is the only aspect of the ESA that may apply to purely private synfuels developments. Generally Section 9 serves to prevent people from shooting bald eagles or importing elephant tusks.

Interagency Cooperation.

The teeth of the ESA, insofar as synfuel and other energy projects are concerned, are found in Section 7 of the Act, innocuously entitled "Interagency Cooperation." In pertinent part, it requires:

> All other Federal departments shall, in consultation with and with the assistance of the Secretary, utilize their authorities in furtherance of the purposes of this chapter by carrying out programs for the conservation of endangered species listed pursuant to Section 1533 of this title and by taking such action necessary to insure that *actions authorized, funded or carried out by them do not jeopardize the continued existence of such endangered species and threatened species or result in the modification of habitat* of such species which is determined by the Secretary, after consultation as appropriate with the affected States, to be critical.

Section 7 imposes two requirements on a federal agency. It must study the impact that an action it authorizes, funds, or carries out will have on endangered or threatened species. And it must consult with the Department of the Interior on how best to minimize any adverse impacts. Regulations promulgated by Interior require a minimum 90-day endangered species consultation period. Any agency seeking to go forward with an action before doing either of these two things can be stopped by a court order, and Section 11 of the law provides that citizens may sue to enforce the ESA.

But the final decision on how to meet the goal of protecting an endangered species lies with the agency taking the action, not with the Secretary of the Interior. The Interior Department has *no formal veto power.* As the leading court decision in the area states:

> once an agency has had meaningful consultation with the Secretary of the Interior concerning actions which may affect an endangered species the final decision of whether or not to proceed with the action lies with the agency itself. Section 7 does not give the Department of the Interior a veto over the actions of other agencies, provided the required consultation has occurred. It follows that after consulting with the Secretary, the Federal agency involved must determine whether it has taken all necessary actions to insure that its actions will not jeopardize the continued existence of an endangered species or destroy or modify habitat critical to the existence of the species.

This does not mean that an agency may disregard the advice of Interior and go forward with a project likely to jeopardize an endangered species. The Federal Administrative Procedure Act generally precludes decisions by federal agencies that are "arbitrary" or "capricious" in view of the evidence before them. Thus, if an agency ignores evidence of threats to an endangered species, particularly if the Department of Interior has urged the agency to change its plan, the decision may be vulnerable to challenge in court.

That is precisely what happened in the famed snail darter case. The Tennessee Valley Authority, which wanted to build the Tellico Dam, consulted with Interior and, in effect, disregarded its advice. Three courts, including the Supreme Court, held that, although the Secretary of Interior could not veto the dam, his findings as to the danger to the snail darter could "properly influence final judicial review of such actions, particularly as to technical matters committed by statute to his special expertise."

Recapping, Section 7 applies only to actions by federal agencies. It imposes a duty to study the effects of a proposed agency actions (such as approving a project) on endangered species and to consult with the Department of the Interior on how to minimize any impact.

Exemptions from the ESA.

During the Tellico Dam-snail darter controversy, Congress passed an amendment to the ESA that created a procedure to *exempt* agency action from the provisions of the ESA where an unresolvable conflict exists between completion of a project and protection of an endangered species. This procedure can be invoked only *after* the study of the impacts and interagency consultation.

The ESA Amendments established an Endangered Species Committee (sometimes called "the God Committee") to hear applications for exemption. The committee is composed of the secretaries of Agriculture, the Army, and the Interior, the Chairman of the Council of Economic Advisers, the Administrators of the Environmental Protection Agency and the National Oceanic and Atmospheric Administration, and, state officials appointed by the President to represent the states affected by a particular project.

The committee may grant exemptions, permitting the possible extinction of a species by federal action, if a majority finds that:

1. There are no reasonable or prudent alternatives.
2. The benefits of the project clearly outweigh the benefits of "alternative courses of action consistent with conserving the species."
3. The project is of national or regional significance.
4. Steps will be taken to mitigate the adverse impacts of the project on the endangered species.

The committee has acted twice. It denied the Tellico Dam an exemption (and was subsequently overruled by Congress), but granted one for the Grayrocks Dam and Reservoir in Nebraska. In the latter case, the developers had established a $7.5 million trust fund to finance specific activities to help save threatened whooping cranes, in addition to other mitigation steps.

Citizen Suits Under the ESA.

Section 11 allows private citizens to sue to enforce provisions of the ESA. Citizens may sue an agency that they believe to have violated the act or sue the Secretary of the Interior to compel him to take action. As with the citizen suit provisions in many other environmental laws, the only limitation on this right to sue is a requirement that the secretary must be given a written notice 60 days before suit may be brought. The law allows the court making a final decision in a citizen suit to award payment of litigation costs, including attorney's fees, to those bringing the suit.

SAFE DRINKING WATER ACT

A second—though somewhat indirect—level of water quality regulation is provided by the Safe Drinking Water Act. This law empowers the Environmental Protection Agency to establish health standards to protect public drinking water sources, including aquifers with less than 10,000 milligrams per liter of total dissolved solids, which, though not now drinking water sources, could become so in the future.

Under the act, EPA has promulgated a set of National Interim Primary Drinking Water Standards for 17 chemicals, organic turbidity, coliform bacteria, and radionuclides. The standards establish the maximum contaminant levels (MCLs) of these regulated substances which can be allowed in drinking water. EPA has also proposed, but not finalized, National Secondary Drinking Water Standards. Secondary standards are not mandatory, and deal with contaminants that affect the odor or appearance of the water in such a way that persons served by a system exceeding MCLs of the secondary substances might discontinue using the water.

Enforcement of the act is given to states assuming primary enforcement responsibility. To do so, each state must adopt drinking water standards no less stringent than the federal primary standards and implement procedures to enforce those standards. The secondary standards are guidelines only.

SDWA regulations, as adopted by the states, are aimed primarily at publicly-owned water systems, rather than generators of industrial pollution. There are, however, several ways in which the SDWA does impact industrial polluters and, potentially, synfuel plants.

First, the SDWA primary standards affect *water quality standards* promulgated by states pursuant to the Clean Water Act. Any NPDES permit granted cannot allow effluent discharges which would exceed water quality standards.

Second, SDWA requires that states develop a program to regulate underground injection: the injection of liquids into or near underground water-bearing stratum. Some synfuel projects—notably shale projects—have explored the notion of reinjecting mine wastes and other process waters into underground cavities in order to minimize surface discharges. If reinjection is into non-drinkable aquifers, such as the Saline Zone of the Piceance Basin, and has no effect on other underground water supplies, this regulatory program will not apply, but where that is not the case the injection operator will be required to get a permit before any discharges can take place.

Third, the SDWA standards have an impact on solid waste disposal regulations under the Resource Conservation and Recovery Act (RCRA). That act, discussed in a previous section, places particularly stringent requirements on the disposal of hazardous wastes. Among wastes automatically classified as hazardous are those whose leachates contain 100 times any of the primary drinking water standards. Based on this criterion, wastes from several steps in conventional oil refining have been classified as hazardous—due to the presence of lead or chromium. It is probable that similar processes in synfuel refining or upgrading facilities would also generate hazardous wastes as now defined.

STATE SITING LAWS AND PROCEDURES

Every synfuel project will require state approval for some permits and other actions before the plant can be built or operated. Some states administer the permitting programs under the major federal laws which we have been discussing; a PSD or NPDES permit will be approved by state authorities. All states have requirements that are not mirrored in federal laws: a plant will require decisions on zoning, construction permits, rights to cross or close state roads and so on. Non-hazardous solid waste disposal remains the province of state regulation. A few states require that major energy facilities receive special permits before they begin construction.

Most of the states with large coal or shale reserves have developed procedures to streamline, speed up, or centralize state actions on a major project with significant environmental impacts. We will refer to these procedures as State Siting Laws.

Three approaches stand out:

A state may hold *joint hearings* or *review*. States that do this, such as Colorado and Utah, keep existing regulatory authority where it is—the water authority handles water permits, the air authority, air permits, etc.—but develop a master schedule under which the applicant presents information relevant to all permits and approvals in a single process and receives approvals, denials, or modifications during that process.

Some states have gone beyond this to adopt *one-stop permitting,* where a single government body is given the authority to issue all permits associated with certain kinds of industrial facilities.

And some states have passed *state facility siting laws* that require that a high-level state authority decide whether the state even wants to have the facility located before any further activity takes place. In theory, an applicant who could meet the requirements for each and every permit could still be told "no." In practice, in states such as Wyoming and Montana, siting laws are used to "encourage" applicants to mitigate socioeconomic impacts and to develop a broad overview of the impacts of a project before approving it part-by-part.

These three approaches do not exhaust the possibilities of state action. Many states have also established *statutory decision deadlines* (like those which exist for federal PSD permits) which guarantee that decisions will be made within a certain period of time after permits are applied for. Some undertake computer tracking of where in the permit maze a particular project is and where it is going. And some appoint *permit coordinators* to work with applicants in shepherding the project through the permitting process.

It is important for you to know how your state handles permitting, so that you are aware of when decisions affecting the aspects of a proposed project that you find most troubling will be made. Even though the synfuels industry is barely off the ground, citizen groups have already discovered to their dismay that important decisions have been made before they were prepared to participate, often before they even knew that the issue was before any state official for resolution.

When a state consolidates permits, conducts one-stop permitting, or requires a preliminary facility siting permit before a project can go forward, citizens have a golden opportunity to participate and make their views known. Like the federal air, water, and waste permits, these state proceedings are often run by rules which work to the advantage of concerned citizens. Generally, the applicant is required to make a case in support of the project. Hearings are open, public and conducted with common sense rules for presenting evidence. Often citizens are given the opportunity to question the project proponents, exposing aspects of a project that are not heralded in press releases.

In some states, the "state siting hearing" may provide the best chance for citizens to stop or dramatically modify a proposed project. The following material will give a brief description of state siting laws and procedures in the states most likely to be faced with synfuel development. It is not intended as a primer on how to participate in state proceedings, but as a guide, to let you know in general terms how your state operates when it faces a decision on a synfuel plant.

Colorado

Colorado has established the Colorado Joint Review proccss (JRP) for Major Energy and Mineral Resource Development Projects. The JRP is totally voluntary. Each project elects whether or not it wishes to participate. The state then decides whether the project is appropriate for the JRP.

The process works as follows. After a project applies for Joint Review status, the Colorado Department of Natural Resources (DNR) decides, in conjunction with other state agencies, whether or not to accept the project. If the project is accepted—and none have been rejected—the state organizes and schedules the activities involved in permitting. One key element of the organizational phase is developing contacts with the federal and local agencies that review permits. The state designates a lead agency, usually DNR, to make contacts with decision makers from other levels of government.

The goal is for all permitting authorities to agree on a single project decision schedule, governing the completion of relevant action by the project and government agencies. The agreement is not legally binding, but does represent a commitment by the agencies to attempt to meet the schedule.

During the organizational phase, DNR holds at least one or two public meetings on the project. The meetings give members of the public a chance to voice concerns, as well as an oportunity to learn more about the proposed project. But comments made by citizens at such meetings do not necessarily trigger responses from any of the permitting agencies. If you are concerned about water pollution, you must still participate in the NPDES permit process.

After the organizational stage is finished and a master schedule prepared, the joint review team monitors implementation of the schedule. The master schedule may provide for joint hearings, if they are otherwise legal, but it creates no changes in the procedures which must be employed for each and every permit.

The master schedule can be helpful to citizens, because it shows what permits a project must obtain, who decides on the application and approximately when and where hearings will be held and decisions made. It provides an overview of applicable legal requirements, but does not change them or add new requirements.

In addition to the JRP, Colorado has established short deadlines for action on state air and water permits. Decisions on state air permits must be made within 135 days of the filing of a complete application; water permits must be granted or denied within 180 days of filing of a complete application. Colorado currently administers the NPDES program, but does not administer the PSD program.

Illinois

Illinois has established, by Governor's Executive Order, a Coordinated Review Process for major non-nuclear energy projects. The Illinois process is modelled on the Colorado JRP.

Participation in the review process is optional. A project may apply to the Illinois Institute of Natural Resources. INR examines the project and makes a recommendation to the state Energy Review Board, composed of representatives of seven state agencies, which in turn makes a recommendation to the Governor. If the Governor approves, all state agencies with permitting responsibility must participate. Federal and local agencies with permitting authority are invited to participate as well.

For each project, a review team, consisting of the project sponsor and one lead agency from the federal, state and local government, is established. The team establishes a review schedule and conducts public meetings.

Illinois has also developed a Consolidated Permit Review which applies to any project seeking permits from more than one branch of the Illinois Environmental Protection Agency. The IEPA has divisions dealing with air, water, land, and public water supplies. The IEPA administers both the NPDES and PSD programs in Illinois, as well as SDWA programs.

Participation in the Consolidated Permit Review is optional, but if the sponsor participates IEPA requires that it submit a "total project application." Each division of IEPA retains authority over its permits, but public hearings deal with the total environmental impact of the project.

At the end of the process, IEPA either issues all of the requested permits or none of them. Consolidated Permit Review can, of course, be incorporated into the larger Coordinated Review Process.

Indiana

Indiana has no programs to consolidate or streamline energy facility permitting. Water permits are handled by the Stream Pollution Control Board. Air permits are granted by the Air Pollution Control Board. And the Environmental Management Board handles solid waste permits.

Indiana administers both the NPDES and the PSD programs through its state agencies.

Kentucky

Kentucky is just beginning to implement a "priority application procedure" for projects that require multiple permits. Like the Colorado and Illinois programs, Kentucky's program involves administrative changes, rather than a statutory reorganization of permitting authority.

The priority application procedure is administered by the Kentucky Department of Natural Resources and Environmental Protection. The Secretary has authority to designate a project a priority application. If he does so, each state agency issuing environmental permits must attempt to expedite decisionmaking on the permits. An interdepartmental task force coordinates state activities.

Kentucky currently does not administer the NPDES program within the state. It has received a full delegation of PSD authority from EPA Region IV. This means that it handles both technical and administrative aspects of air quality permit issuance, pursuant to EPA's regulations.

Montana

Montana Major Facilities Siting Act establishes a coordinated review process for major energy facilities. The process employs a master application and mandates consolidated hearings, but stops short of providing "one-stop" permitting. In fact, the process can best be understood as two-stop permitting under strict decision making deadlines.

The Facilities Siting Act requires that any synfuel project receive a Certificate of Environmental Compatibility and Public Need from the Montana Board of Natural Resources. While the Board makes the final decision on the permit, the Department of Natural Resources serves as its staff and conducts hearings and reviews. At the same time that BNR and DNR are conducting their review of the application for the certificate, the Montana Department of Health and Environmental Services reviews the project's applications for state air, water, solid waste, and hazardous waste permits. DHES makes its decisions on these permits independently of BNR, but still follows a timetable established by the siting act.

The process begins with the filing of a master application with DNR. When the application is declared complete, DHES begins to process permits, while DNR prepares a Montana Environmental Policy Act report.

DHES must make decisions on all permit applications within one year. Any permit decision may be appealed to the Board of Health, which must rule on them within 18 months of the date an application is declared complete.

In the meantime DNR prepares a draft environmental assessment and, where needed, environmental impact statement (Montana has a state NEPA). It must hold public hearings. Within 22 months of the completed application, DNR must prepare recommendations to BNR on the application.

Citizens, state and local agencies, and the project sponsors may contest DNR recommendations in hearings before BNR. BNR has 11 months to rule on the DNR application and issue or deny the certificate.

The board has the authority to place conditions on the granting of the certificate beyond those required for specific permits. In one case, BNR required a power company to set up a training program for members of an Indian tribe wishing employment on the facility receiving the permit.

Because Montana administers the NPDES program, all water permits are covered during the siting act process. Montana does not administer the PSD program.

New Mexico

New Mexico has no programs for coordinating state permitting. The state has established some statutory deadlines for important permits. Groundwater discharge permits must be approved or denied within 60 days of the submission of a complete application. State air quality permits must be decided on within 30 days of submission of a completed application.

New Mexico does not administer either the NPDES or PSD programs, but EPA Region VI expects to make a full delegation of PSD authority sometime in 1982.

North Dakota

North Dakota has a state siting law that requires major energy facilities to be approved by the Public Service Commission. While the PSC decides whether a project will receive a "certificate of public need and convenience," it is not the only state agency involved in issuing permits. Potential applicants must notify the PSC one year before filing their application.

The PSC may not begin consideration of the application until the State Engineer or Water Commission (depending on the amount of water required) issues a water appropriation certificate. At that point, the PSC takes the siting application under advisement. It must hold a public hearing before reaching any decision.

All state environmental permits are handled by the North Dakota Department of Health (except for mining permits). The state administers PSD, NPDES, RCRA, and SDWA programs. The Department of Health schedules a hearing on a PSD application only if it receives a request to do so.

Ohio

Ohio has a one-stop permitting process for electric generating facilities and transmission lines, but the process does not currently apply to synfuel plants. Permitting for synfuels will be done through the ordinary state permitting procedure.

Most permits are handled by the Ohio Environmental Protection Agency (OEPA). The Water and Land Pollution Control Office of OEPA has five regional offices, and the Office of Air Pollution Control has sixteen. Project sponsors submit applications directly to these field offices. The Office of Air Pollution Control has the authority to issue combined air and water permits.

Ohio administers the NPDES program and has received full delegation of the PSD program from EPA Region V.

Pennsylvania

In Pennsylavnia, all environmental permits are issued by the Department of Environmental Resources (DER). The Department has seven regional offices. The director of the regional office or his designee may coordinate environmental permit processing for any facility located in his region.

If an applicant chooses to use a regional office to coordinate permitting, then all correspondence and materials relating to the project are channeled through the regional office to the bureaus which actually grant or deny permits. The applicant still must complete each permit application and retains the option of dealing directly with the bureaus rather than going through the regional office.

Pennsylvania administers the NPDES program. It has received partial delegation of PSD program. The state undertakes technical analysis and processing of PSD applications, but final decisions are still made by EPA Region III.

Tennessee

Tennessee recently enacted legislation to etablish a joint review process and set decision making deadlines. The Major Energy Project Act establishes procedures for joint review of large energy projects.

Potential applicants must apply to the State Department of Economic and Community Development for designation as a priority project. Final designation authority rests with the governor. All state and local agencies are required to join in the joint review process; federal agencies are invited.

A joint review team is established including state agencies designated by the Governor, local agencies, and federal agencies which choose to participate. The act requires that all permitting action be completed within two years. This provision is binding on state and local agencies.

The joint review team develops a project decision schedule. If a state or local agency fails to comply, the team leader may bring an enforcement action in state court.

The act gives the review team the authority to waive "any state or local statute, regulation or requirement" enacted after the commencement of construction of a facility if necessary to ensure timely and cost-effective completion of a facility without the endangerment of public health or safety. Commencement of construction occurs as soon as binding contractual commitments are made or on-site construction begins, whichever is earlier. Tennessee administers both NPDES and PSD programs.

Utah

In 1981, Utah passed legislation to establish a joint review process. The legislation created a Resource Development Coordinating Committee in the State Planning Office to oversee the review process. The Committee consists of 20 representatives of state agencies plus ex officio members from the Bureau of Land Management, National Park Service, and U.S. Forest Service.

Participation in the coordinated review process is voluntary. The review team establishes a non-binding schedule; it is also responsible for scheduling public meetings and hearings on the project.

The legislation did not change any regulatory responsibilities or establish decisionmaking deadlines. Utah does not administer the NPDES program, but does administer PSD.

Virginia

The administrator of the Virginia Council on the Environment has some statutory authority to coordinate services for applicants seeking multiple permits, but this authority has not been extensively utilized in the 12 years since the Council was established. The council consists of the chairmen of the eight state agencies primarily responsible for environmental protection, two gubernatorial appointees, a chairman and an administrator.

The administrator may publish combined public notices, conduct consolidated public hearings, or coordinate the issuance of permit decisions, but this has rarely been done.

Virginia administers NPDES and has received full delegation from EPA Region III for the PSD program.

West Virginia

The Governor's Office of Economic and Community Development serves as a permit coordinator in West Virginia, on request. The applicant must still apply separately to each state agency from which he needs a permit.

West Virginia does not administer the NPDES program, but it has received full delegation from EPA Region III for the PSD program.

Wyoming

Major energy projects in Wyoming must be certified by the Wyoming Industrial Siting Council. The council was established by the Wyoming Industrial Development Information and Siting Act of 1975. Its primary objective is to make developers predict social and economic impacts stemming from large energy projects and commit themselves to monitoring and mitigation strategies.

Although environmental issues are relevant to the council's deliberation, the act is primarily geared to giving the state the best possible leverage to control boomtown development.

As with the North Dakota siting act, the Wyoming law requires a preliminary showing that the applicant has provided for its water needs. Certification by the State Engineer is required before the council can begin its public hearings on the project.

The law sets no deadlines for permit decision making, nor does it permit or require consolidated hearings or joint review of environmental permits. Wyoming administers both the NPDES and PSD programs.

8

Other Useful Laws for Citizens

A myriad of laws affect synfuels other than those that regulate environmental effects. In this chapter we discuss three federal laws that are particularly important to citizens.

THE NATIONAL ENVIRONMENTAL POLICY ACT (NEPA)

We separate NEPA from the seven federal laws discussed in the previous chapter, because it is the only purely procedural environmental law of consequence. No provision of NEPA regulates stack heights or limits emissions. No rulemaking under its authority prescribes best management practices or tells a miner how to place spoil on the downslope. In fact,

NEPA does not apply to private individuals or corporations at all. It applies only to *federal agencies undertaking major federal actions affecting the environment.*

There are three reasons why NEPA is so important in fights over energy development. It is a *stop sign*: nothing final can happen until it has been complied with. It is an *information process:* NEPA, properly used, lays bare enormous emounts of information about a proposed project. This information may not only have a bearing on the particular decision being made by the federal agency preparing an environmental impact statement, but also may affect many other decision makers, including political and financial interests who make decisions that are not constrained by regulations. Finally, NEPA is *cheap and easy to use.* Anyone who can speak or write can participate. The legal procedures are relatively simple.

What NEPA Does

The key to NEPA is found in Section 102(2) (C) of the act, which quite simply states that:

> all agencies of the Federal Government shall include in every recommendation or report on proposals for legislation and other major Federal actions significantly affecting the quality of the human environment, a detailed statement by the responsible official on—
>
> (i) the environmental impact of the proposed action,
>
> (ii) any adverse environmental effects which cannot be avoided should the proposal be implemented,
>
> (iii) alternatives to the proposed action,
>
> (iv) the relationship between local short-term uses of man's environment and the maintenance and enhancement of long-term productivity, and
>
> (v) any irreversible and irretrievable commitments of resources which would be involved in the proposed action should it be implemented.
>
> Prior to making any detailed statement, the responsible Federal official shall consult with and obtain the comments of any Federal agency which has jurisdiction by law or special expertise with respect to any environmental impact involved. Copies of such statement and the comments and views of the appropriate Federal, State and local agencies, which are authorized to develop and enforce environmental standards, shall be made available to the President, the Council on Environmental Quality, and to the public as provided by section 552 of title 5, United States Code [Freedom of Information Act], and shall accompany the proposal through the existing agency review process.

In most instances, the "report" described in Section 102(2) (C) is called an *environmental impact statement* (EIS). Sometimes, under conditions to be outlined below, a federal agency may prepare an *environmental assessment* (EA) or make a *finding of no significant impact* (FONSI). These are both shortcuts in the NEPA process and, while they may be appropriate in certain cases, they should always arouse some suspicion.

In the rest of this chapter we will walk through a hypothetical NEPA proceeding. The Council on Environmental Quality has issued regulations which act as guidelines to all agencies, describing certain procedures which must be followed, but each agency has its own regulations defining its NEPA process.

When Is an EIS Required?

The preparation of an EIS is only required for a federal action which is a "major federal action significantly affecting the quality of the human environment."

At the outset, a federal action requires that the decision be made by a federal agency. Actions by state and local agencies under federal laws are not federal actions. Federal participation need not be the dominant factor, however. A joint private and federal project is subject to NEPA, even if federal participation amounts to less than private participation.

Exemptions. Legislation and judicial decisions have created several exemptions to NEPA by declaring that certain federal agency actions are *not* major federal actions for the purposes of NEPA. These exemptions do not remove projects totally from NEPA. What they do is to say that a given agency action—which for all the world appears to be a major federal action— will not trigger NEPA responsibilities for that agency.

An example might be helpful. The Energy Security Act states that funding decisions by the Synthetic Fuels Corporation will not be considered as major federal actions (except in the unlikely case that the SFC puts up all the money to build a Government-Owned, Contractor-Operated (GOCO) facility). This does not mean that any synfuels project which gets SFC money is exempt from NEPA. It means that the SFC does not have to worry about preparing an EIS based on its funding action.

Similarly, many permitting decisions made by the Environmental Protection Agency are also considered not to be major federal actions. Permitting decisions under the Clean Air Act, particularly the Prevention of Significant Deterioration permit, do not trigger EIS preparation by EPA. A more complex rule applies to permits issued under the Clean Water Act. The granting of an NPDES permit is considered a major federal action only if the facility is classified as a *new source.* A new source is a facility for which EPA has issued a standard of performance. At present, a coal liquefaction, gasification, or oil shale facility would not be considered a new source, because EPA has not issued a standard of performance. But it has done so for coal mines. So a large synfuel project which included a coal mine would be considered a new source and its application for an NPDES permit from EPA would trigger NEPA. But the NPDES permit for a project which does not contain a coal mine is not a major federal action.

Types of Federal Actions. Almost anything that a federal agency does involving a synfuels project is potentially a major federal action, unless it has been specifically exempted. Among the more common forms of federal action are funding (through contracts, grants, subsidies, or loans), leases, permits, licenses, or certificates.

The making, modification, or establishment of rules, regulations, procedures, or policy may also be federal actions.

When Are Actions Major?

Under the provisions of NEPA, an agency can avoid preparing an EIS on a non-exempt federal action, if it determines that the action is not major or that it does not affect the environment significantly. There has been extensive litigation on the question of whether particular actions are major actions that significantly affect the human environment. It is difficult to specify which actions will generally invoke the application of NEPA, but the courts have generally interpreted the term "major action" quite liberally.

And NEPA does not just allow an agency to walk away from its responsibilities with a shrug and vague statement that nothing major is happening. A specific procedure governs the "no-action" finding. If the agency rules do not already specify that a non-exempt action is either definitely a major action or always not a major action, it must prepare an *environmental assessment* (EA), a brief examination and analysis of the proposed action, and make it available to the public for comment. If it determines on the basis of an EA that the action is not major or that the impact on the environment is not significant, it must prepare a second public document called a *finding of no significant impact* (FONSI) setting out its reasons for that decision.

A finding of no significant impact is a final agency decision and therefore subject to court challenge. Although the EA-FONSI process does not provide the full opportunity for public review that is found in the EIS process, it does allow public review and participation, making it difficult for agencies to "pull a fast one."

How Soon is an EIS Required?

The CEQ rules and a body of court decisions encourage preparation of an EIS as early as possible. Practical reasons support this approach: it is difficult to select options after concrete has been poured and millions of dollars spent. But an EIS can be prepared too early—before a project is in its final form or critical information gathered. It has not been unusual thus far in the synfuels program to see EIS analysis of projects where the sponsors have yet to make key process or siting choices. This robs the process of the opportunity to scrutinize and, where necessary, change plans.

Sometimes when there have been major changes in plans for a project or in regulations governing a certain type of activity, *supplemental* statements are prepared. This is rare and one should not hold one's breath waiting for it to happen.

When the first public notice of intent to prepare an environmental impact statement appears, the process is already well in gear. In most cases, the applicant or project will already have been in contact with the federal agency concerning its lease, license, contract, or funding. In many instances, the applicant will already have furnished the agency with its evaluation of the environmental impact of the project—often called an Environmental Information Document, or EID. And in some cases, where the applicability of NEPA is in question, the agency will have prepared an Environmental Assessment which leads it to conclude that there would be a significant impact.

All this material can be immensely useful to citizens concerned about a potential project, particularly if they plan to make a major effort during the EIS process or if they are involved in some of the early permits which are not covered by NEPA (state permits, PSD, Synfuels Corporation funding decisions). The Freedom of Information Act can and should be used to obtain much of this material.

What is Covered by an EIS?

Regardless of what major federal action triggered an EIS, *the whole project can be examined.* You might have a Bureau of Land Management right-of-way for a two-mile rail spur as the major federal action, but that does not limit the EIS. It can cover the entire project, from mining the resource to disposing of the wastes. It can look at areas that BLM has no regulatory authority over, that no one in the federal government can regulate. It can cover the interaction between the project under review and other projects. Within reason, the scope of an EIS can be extremely broad.

Just as important, *the EIS must examine alternatives* to any proposed action, including the alternative of doing nothing. This is extremely important. Assume that a proposed plant will locate on a floodplain, or will send 10 unit trains through the center of town every day, or does not employ the best pollution control system. The EIS can examine the obvious alternatives and compare them to the proposal, but only if those alternatives are chosen for study during the scoping process.

Scoping is the first formal step in the EIS process and will be discussed in the next section. But a good deal of informal activity goes on before scoping and it helps to have followed it as closely as possible. Know as much about the project as possible. Billion dollar projects do not spring up out of nowhere. There will have been public and corporate decisions. Land and resources will have been acquired. Initial choices of technology and contractors will have been made. The project may have already applied for some permits not covered by NEPA or applied for funding from the Synfuels Corporation. Most likely the project sponsors will have come to the location to explain their project, the benefits it will bring, and the steps they will take to deal with potential problems. And in most instances they will have filed some equivalent to an EID with the agency preparing the EIS. A lot of information will be available—to those who will dig a little to get it.

Scoping

The real first step in the formal EIS process will be a rather vague notice in the Federal Register announcing the intent to prepare an environmental impact statement for a project. For those who do not read the Register every morning, news about the project and the EIS will probably be disseminated through the local news media, but once the arrival of a project becomes imminent it is helpful to start following the Register. Libraries carry it. Since it may not be obvious which agency will prepare the EIS, you may have to scan the Table of Contents. Also check in the section at the end of the Table, entitled Meetings Announced in This Issue. Look for the announcement of a scoping meeting.

The scoping meeting marks the first local appearance of officials from the agency which will prepare the EIS. The agency does not have to hold an actual meeting; it can just ask for written suggestions; but a strongly worded letter, a call to one's Congressman, and other forms of pressure will probably produce a public meeting. There are several reasons to seek a public meeting. It gives you a chance to meet some of the officials who will be in charge of the EIS and gives them a chance to meet you: that makes phone calls and exchanges of information later a lot easier. It lets concerns be expressed verbally, which is often easier and a much better way to demonstrate intensity of concern. It provides a forum in which the process—both the EIS and the federal actions which are contemplated—can be explained and you can ask questions. These explanations are almost always superior to receiving copies of dry and obtuse memoranda and regulations. And, of course, any public meeting can be a source of marvellously free publicity for your views and concerns.

There is some specific business which should be completed at any scoping meeting. It is important to find out what documents have already been prepared and how to get them. The scoping process often sets a timetable for completing the draft and final statements: make sure it is one you can live with.

But the most important issues to be dealt with at the scoping meeting are decisions about what elements of the proposed projects will be studied and what alternatives will be examined. In general, you will want as much of the proposed project studied as possible, and that is easy to ask for. Choosing an alternative is far more difficult. Knowing what your major concerns with the proposed project are ahead of time is critical. All too often, an EIS will contain a preferred alternative (what the agency or applicant wants to do) and other alternatives that are obviously impractical. If you have a specific problem with the proposed project and a specific solution, make sure that it gets analyzed. The items you might want to pay most attention to are location of the plant, pollution controls, source of supply method, and location of transportation.

Interagency Consultation

A series of other important laws requires that federal agencies make sure that their activities are in compliance with standards established by those laws or that agencies about to take actions which could have an impact on other values consult with those federal agencies charged with protecting those values. We discussed one such law earlier, the Endangered Species Act.

The purpose of such compliance and consultation provisions is to make sure that the agency taking the federal action is aware of the potential consequences of its actions and to provide a method for working out potential conflicts. The EIS process is an obvious place for such compliance reviews, consultation, and interaction to take place. So most agencies, in developing internal guidelines for NEPA procedures, specify just which consultations and compliance reviews are mandatory, as part of the NEPA process.

In addition to the Endangered Species Act consultation, it is likely that the followng consultations will take place during the NEPA process:

> National Historic Preservation Act consultation. This Act, along with the Archaeological and Historic Preservation Act of 1974 and Executive Order 11593, establishes a review, comment, and mitigation process with respect to any federal action which affects any property with historic, architectural, archaeological, or cultural value. The Advisory Committee on Historic Preservation is afforded an opportunity to comment on any action which affects properties listed above. To meet the consultation requirement most agencies will prepare a historical and cultural resource survey.
>
> Fish and Wildlife Coordination Act consultation. This Act requires Federal agencies involved in projects which might adversely affect fish and

wildlife to consult with the Secretary of the Interior and State Wildlife Agencies. In some instances, a special field investigation by the Fish and Wildlife Service may be necessary. Where some adverse impact is found, the Fish and Wildlife Service or the State Wildlife Agency may recommend that mitigating activities be undertaken. Mitigation may inclued a modification in the project, a modification in the way the project operates, or the provision of some offsetting benefit.

Executive Order 11988 (Floodplain Management) and Executive Order 11990 (Protection of Wetlands). These orders require that every federal agency adopt regulations ensuring that the potential effects of any action they may take in a floodplain or wetlands are evaluated and minimized. Most agencies have incorporated this into the EIS process, where an EIS is required.

The Wild and Scenic Rivers Act. The agency must determine whether the action will affect a segment of any river which has been designated pursuant to the Act or is under study. Depending on the classification of the river and the significance of the impact, the Secretary (Interior or Agriculture depending on the management system in which the river is located) may allow certain activities. The Act, however, bars certain other types of activity.

Synfuel projects funded or authorized under the Federal Non-Nuclear Research and Development Act must be studied by the Water Resources Council to determine the impact of the project on water use and availability.

The compliance and consultation process also normally includes a review of the compliance of the proposed project with the permit requirements of the Clean Air Act, Clean Water Act, RCRA, and the Safe Drinking Water Act. (Remember: the issuance of these permits is often not a major federal action.)

The Draft EIS

When all preliminary analysis of the impacts of the project are complete, the agency will prepare a draft environmental impact statement. The bulk of the work on the DEIS is usually done by outside contractors, but the agency responsible for the EIS must examine it, revise it where appropriate, and present it as its own.

The Draft EIS (or DEIS) must contain:

—A general description of the project;
—A discussion of the purpose and need for the project;
—A description of alternatives studied;
—A description of the existing environment;

—An analysis of the environmental consequences of the alternatives under study;
—A description of federal, state, and public participation in preparing the draft;
—A list of those who prepared the draft;
—A bibliography;
—An index; and
—Appendices.

The Draft EIS is a public document. Its publication is reported in the Federal Register and it is made available to the public free of charge. The Draft must normally be made available at least 30 days before any public hearing is held on it.

Public Comment

Once the Draft EIS has been made public, there is a period of time (normally 90 days) in which the public and other agencies may comment on it.

Comments should be specific and documented. They should explain the issue, the reason that it is important, and offer documentation and authorities in support of the comment.

Comments may be in writing—and the most technical should be—but there is usually at least one public hearing to receive comments on a controversial DEIS. This public hearing is likely to be somewhat more formal than the scoping hearings. A transcript is made. Often it will be necessary to sign up in advance in order to speak, and there may be time limits.

The public hearing can be a media event, providing an opportunity to make the community and government officials aware of issues raised by the proposed plant.

Utilizing the resources of local universities and colleges, as well as the services of concerned citizens with professional abilities, local groups have been able to prepare detailed and devastating critiques of shoddily prepared Draft EISs. Detailed comments can educate the community and its politicians and can lay the groundwork for a challenge to any final agency decision.

Final EIS

After the comment period on the Draft EIS closes, the agency must prepare a final environmental impact statement (FEIS).

The CEQ rules indicate that the preparing agency must assess and consider comments both individually and collectively. It is not enough merely to indicate disagreement with a comment or critique, as one court stated:

> officials must give more than cursory consideration to the suggestions and comments of the public in the preparation of the final impact statement. The proper response to comments which are both relevant and reasonable is to either conduct the research necessary to provide satisfactory answers or to refer to those places in the impact statement which provide them. If the final impact statement fails substantially to do so, it will not meet the statutory requirements.

Lathan v. *Volpe,* 350 F.Supp. 262, 265 (W.D.Wash., 1972).

Record of Decision

The Final EIS is an analysis. While it may contain a preferred alternative, it is not, in itself, an agency decision, license, grant, lease, etc. That must be made in accordance with the procedures established by whatever law gives the agency the power to take that action.

When the agency makes its final decision on the proposed action, it must include a *record of decision.* The record of decision is an account of how the material contained in the EIS influenced the agency's choice among various options and any mitigating measures which the agency will require.

The filing of the record of decision is the last step in the formal agency EIS process. Once the record has been prepared, then the ban which NEPA placed on agency action before environmental analysis is over. That is, unless there is a NEPA lawsuit.

NEPA and Judicial Review

Agency compliance with the procedural requirements established by NEPA is subject to judicial scrutiny. Although the Act does not have a provision specifically authorizing judicial review or enumerating judicial remedies, the courts have uniformly held that *injunctive relief* will be granted to plaintiffs who demonstrate that the requirements of the act have not been met. The courts have been increasingly reluctant in recent years to find violations of NEPA, however.

Once an agency has complied with NEPA's procedural requirements, NEPA loses its sting. The Supreme Court has held that NEPA does not require an agency to give priority in making its decisions to environmental considerations. In the case of *Vermont Yankee Nuclear Power Corp.* v. *Natural Resources Defense Council,* it held:

> NEPA does set forth significant substantive goals for the nation, but its mandate to the agencies is essentially procedural. It is to insure a fully informed and well considered decision, not necessarily a decision the judges of the Court of Appeals or this Court would have reached had they been members of the decision-making unit of the agency. Administrative decisions should be set aside in this context, as in every other, only for substantial procedural or substantive reasons as mandated by the statute.

In other words, an agency can legally come up with a procedurally correct, environmentally horrible decision and not be in violation with NEPA. So does this all leave NEPA as merely a hollow shell? Hardly. The requirement for a thorough, public environmental analysis is extremely important. It has served to improve many an environmentally harmful proposal. And even if NEPA allows for the possibility of an agency essentially disregarding the result of its analysis, other laws—particularly the Administrative Procedures Act—can partially fill this gap.

State NEPAs

Many states have adopted environmental analysis laws similar to NEPA or have established administrative procedures which seek the same goals. A summary of these state laws is available from the Council on Environmental Quality.

Where both a federal and state environmental statement are required, CEQ has established procedures to minimize duplication.

State NEPA laws may be particularly important in providing for analysis before actions which might be exempt under federal law. For example, in some states which have assumed EPA Clean Air Act permitting, a state NEPA might require analysis before the PSD permit is granted.

Because it comes so early in the process, PSD permit approvals often predate any NEPA review or overlap with it.

FREEDOM OF INFORMATION ACT

The Freedom of Information Act is not an environmental law, but every citizen who deals with the federal government on environmental issues should know what it is and how to use it. The basic tenet of the act is simple: all records and material in the possession of the federal government are public property and must be disclosed on request unless they are specifically exempt from disclosure.

A record is any writing, computer tape, etc., that is in the possession of a federal agency. You may request access to any record that an agency has, but you may not compel an agency to create a record where it does not already exist.

When to Use the Act

You may use the Freedom of Information Act (FOIA) any time you think that a federal agency has records in its possession that will be helpful to you in preparing your case concerning a synfuels plant. The records might be reports by agency scientists, files on a plant or the owners of a plant, records of inspection by agency personnel of a site, information filed by a company seeking a license or permit.

FOIA is extremely easy for the citizen to use, but check first to see if the information you are seeking is not already publicly available. Every federal agency routinely makes material available in agency files, record rooms, and agency publications. Much of the material used to prepare this manual came from such documents; all was made available more quickly and less expensively than if FOIA had been used. So if you want an agency record, first ask if you can have it.

The FOIA Request: Who to Ask

The first hurdle is to determine to whom to direct your formal FOIA request. Knowing whom (which agency or branch) to ask tells you about the procedure you must follow in making the request. You can check agency rules for the Freedom of Information Act in the Code of Federal Regulations. Regulations for some of the more likely agencies to be involved in synfuel matters are listed below:

Agriculture Department(Forest Service)	36 CFR 200
Department of Energy	10 CFR 1004
Environmental Protection Agency	40 CFR 2
Federal Energy Regulatory Commission	18 CFR 3
Interior Department	43 CFR 2
Labor Department	29 CFR 70 (OSHA)
Water Resources Council	18 CFR 701

The regulations will tell you exactly to whom in the agency to send different requests, how to address FOIA requests so that they get maximum attention, the agency's policy on waiving search and copying fees as well as the normal charges for such material and much more.

Generally you can address an FOIA request to:

Freedom of Information Act Officer
[name of federal agency]
[address of federal agency]

The FOIA Request: How to Ask

While each agency has its own specific procedures, you can normally make a request with a simple two-paragraph letter.

The first paragraph need merely read something like:

> Pursuant to the Freedom of Information Act, 5 U.S.C. §552, I request copies of the following records:

The second paragraph should indicate a willingness to pay the agency's standard fees (waivers will be discussed below) and set a maximum amount you are willing to pay. It is a good policy to set a limit of $25-$50, even if you are willing to pay more. The best way to do this is with a line in the second paragraph of your letter which may read as follows:

> I am willing to pay reasonable search and copying fees up to a limit of $_______. If you estimate that total fees will exceed this amount, please notify me before incurring any such charges.

By following a few simple rules, you can make your FOIA request simpler and cheaper and markedly improve the chances of getting what you want. First, be as specific as possible. Requests that read "Give me everything you have on X" will often be returned as vague. If they are not, you may end up paying for a lot of search time and copies of materials that you do not need.

Second, consider the option of not having materials copied. You may request the right "to inspect" records. That means the agency will have to make the files available to you in a public reading room, where you can examine them. If you see something important that you want to copy, you can. This approach is particularly useful if you think you will have to wade through mountains of documents in hopes of finding small nuggets of information.

After the Request

According to the law and all regulations, you should receive a reply to your request within 10 working days of its receipt. Never plan on getting a reply in that short a time. FOIA offices are notoriously understaffed. If you really need something by a certain date, make sure that you file your request months in advance.

The best tack to take with FOIA officers concerning when requested material should be available is to be firm, but not obnoxious. You will be given the phone number of someone assigned to work on your request and your request will be given an identification number. Call that person and inquire about the progress he or she is making. If they will slip the deadline by a couple of days or if your request is complex and large, be reasonable and work something out. Remember, a miffed FOIA officer can decide to do a large and expensive search or just get you off his back by denying the request. (Most FOIA staff are quite dedicated to getting information out, so this is unlikely, unless you push them too far. Be good to them and they are likely to return the favor.)

If your request has not been answered within a reasonable time, then you have the right to treat the failure to respond as a denial and to appeal that denial. Sometimes this is your only option. At any time after the tenth working day, you have the right to treat nonresponse as a denial. But it is normally a good idea to send a second letter setting a date, for instance:

> I am in receipt of a notice from your office that my request for information under the Freedom of Information Act was received on February 1, 1982. The ten working day period for responding to that request ended on February 15, 1982. An additional ten days has passed and I have yet to receive a response. This is to notify you that, if I do not receive a response to my request by March 8, 1982, I shall treat such failure to respond as a denial and appeal that denial.

One reason to wait at least 10 days extra is because the agency always has the option to extend the initial 10 day response for an additional 10 days for "good cause," so you might as well give them the time.

Exemptions and Denials

Unfortunately, despite the obvious intent of FOIA to broaden access to the greatest possible extent, requests are often denied in whole or in part. If the agency denies all or some of your request, they must tell you what they are not giving you, give a statement of the reasons for withholding that information, and inform you of your appeal rights.

Denials are most often based on the nine exemptions contained in the FOIA itself. The exemptions are:

(1) National security information
(2) Internal agency personnel rules
(3) Material specifically exempted from disclosure by other federal statutes
(4) Trade secrets
(5) Internal agency memoranda and policy discussions
(6) Personnel and medical files whose release would constitute an invasion of privacy
(7) Law enforcement records
(8) Information on federally regulated financial institutions
(9) Geological information on oil and gas wells

The exemptions are mandatory only for material covered by Exemption (3). Other exemptions may be waived by the agency. Exemptions (3), (4), and (5) are most likely to be used against you when you seek information relating to synthetic fuels.

It is important to remember that a document or record that includes material exempt from disclosure is not totally exempt. The agency is under an affirmative duty to locate *reasonably segregable portions* of records and make them available to the requester. Therefore, if you request a large document and receive a response claiming that the whole document is exempt, you should challenge that contention. At least part of most documents is discloseable.

Litigation defining the scope of the nine exemptions under FOIA has been extensive, and it is not the goal of this manual to serve as a guide to the case law on the Act. However, a brief discussion of the three key exemptions likely to be applied against you may enable you to judge when an agency denial is dubious and when it is plain wrong.

Exemption (3): Other Statutes. Exemption (3) provides that FOIA does not apply to matters that are:

> specifically exempted from disclosure by statute (other than section 552(b) of this title) provided that such statute (A) requires that the matters be withheld from the public in such a manner as to leave no discretion on the issue, or (B) establishes particular criteria for withholding or refers to particular types of matters to be withheld.

In other words, the statute has to say "This information can not be released, ever," or something to that effect for Exemption (3) to apply. Thc classic examples are laws involving tax returns, CIA functions, grand jury secrecy, patent applications, etc. Statutes that protect confidentiality or business data are generally covered by Exemption (4).

Exemption (4): Trade Secrets. Exemption (4) says that FOIA does not apply to:

> trade secrets and commercial or financial information obtained from a person and privileged or confidential.

Unlike Exemption (3), Exemption (4) is not mandatory. But when trade secrets and business information are involved in an FOIA request, agency personnel will often get very nervous and take you aside, telling you that if they release this kind of information they could go to jail under a federal criminal statute 18 U.S.C. 1905, the Trade Secrets Act. Don't buy it; that law has been on the books for 60 years and no one has even come close to going to jail.

The way the courts have interpreted the exemption, material is exempt if it is either (1) a trade secret, or (2) information which is (a) commercial or financial, (b) obtained from a person, and (c) privileged or confidential.

What is a trade secret? In FOIA litigation, it has been defined as:

> An unpatented, secret, commercially valuable plan, appliance, formula, or process, which is used for the making, preparing, compounding, treating, or processing of articles or materials which are trade commodities.

What about the second half of the exemption? Financial data is rarely a concern in assessing the environmental impacts of a synfuels facility. (If an industry wants to assert that the application of certain standards is too costly, then it is up to them to provide the data, financial or otherwise, to make their case.) But much that goes on in a synfuels facility can be classified as commercial information without stretching the term too far. Some synfuel developers have sought to shield everything about their plants, including information on pollution and potential health hazards.

The courts have generally protected information concerning profits, losses, market shares, customers, etc., under Exemption (4). Its application to health, safety, and environmental impacts is questionable.

Exemption (5): Agency Memoranda. This is the most widely used exemption and its language is the least clear:

> inter-agency or intra-agency memorandums or letters which would not be available by law to a party other than an agency in litigation with the agency.

Read one way, this exemption could prohibit disclosure of almost every piece of paper generated within the federal government, because almost all of them could be characterized as memoranda. Fortunately, the Supreme Court has held that the purpose of this exemption is to protect "advice, recommendations, and opinions which are part of the deliberative, consultative, decisionmaking processes of government."

Even this definition might present problems. For example, agency evaluations of a permit application play a role in the deliberative process. But the courts have also held that the exemption only applies to "the 'opinion' or 'recommendatory' portion of a document, not to factual information which is contained in the document."

In other words, if an agency report on a permit application begins with a long analysis of the facts involved and then proceeds to make a recommendation based on those facts, the first portion of the document should be available under FOIA, while the second portion might be exempt.

Forcing agencies to segregate nonexempt portions of a record is particularly important when Exemption (5) is involved. Almost every major document subject to this exemption should contain a nonexempt portion. Make sure that you are given access to it.

Appealing a Denial

If all or a portion of your request is denied or if the agency refuses to answer the request at all, you may make an administrative appeal within the agency. You *must* make this appeal if you wish to preserve the option of going to court and suing to overturn the denial. Any denial, whole or partial, should be accompanied by an explanation of that agency's appeal procedures. You must file your appeal within 30 days of being notified of the denial.

The appeal process is disarmingly simple. All you need do is send a letter describing your initial request and the response to it and stating that you are appealing that initial decision. The appeal must be answered within 20 days.

Most agencies release substantially more material on appeal than on initial request.

Your appeal may be just a simple letter. It may contain a legal brief arguing points of law. Or it may argue about the policy reasons that favor release (remember that withholding under any exemption except Exemption (3) is discretionary with the agency). Your appeal letter should ask for a complete list of documents which have been withheld, a description of the documents, and a document-by-document discussion of why material is being withheld, if that has not already been provided.

After the Appeal: FOIA in Court

If the appeal still does not get you the information you desire, then you should consider a lawsuit. All lawsuits involve time and expense. But if you really think the information being withheld is important, look carefully at the option of an FOIA suit. As lawsuits go, they are uncomplicated.

If you win your lawsuit or "substantially prevail," the court may order the government to pay your attorneys fees and other costs. Consider this when you believe you have a strong case or that an agency acted arbitrarily in denying you important information. The court, in deciding whether to award fees, is directed to consider any "public benefit" that comes from the disclosure and when the government acted reasonably in withholding the documents. Never assume you will be awarded attorneys fees, but you can factor the likelihood of this happening into any decision on whether or not to sue.

The Costs of Getting Information

Unfortunately, the Freedom of Information Act is not free, and using it is getting more costly all the time. Agencies may charge for the cost of searching for requested materials and for copying materials. Fee schedules for search and copying are published for each agency in the Code of Federal Regulations. EPA currently charges $2.50 per half-hour for search time and 20 cents a page for copying. Many other agencies have established much higher search fees when professional employees, as opposed to clerical employees, must take part.

A major request can take hours and hours of search time, which is one reason why specific, targeted record requests can be helpful (e.g., letter from John Doe to Joseph Smith, November 19, 1980; or Office of Solid Waste file on RENCO application). No agency can charge you for the costs of reviewing a document to see if they will apply an exemption to it. That is their cost.

Most agencies require that you prepay or make arrangements to pay before they begin their search. Generally, unless extremely large amounts are involved, a simple statement that you are willing to pay up to X amount will suffice. You can be charged for a search that does not yield any documents.

Fee Waivers

The Act provides that agencies may choose to waive fees.

> Documents shall be furnished without charge or at a reduced charge where the agency determines that waiver or reduction of the fee is in the public interest because furnishing the information can be considered as primarily benefiting the general public.

This is not a guarantee that you will get a waiver, particularly if you are seeking information to contest an agency permit decision later.

Most agencies have established regulations which deal with processing requests for fee waivers. You should make such a request in your initial FOIA request. If your request for a waiver is denied, you can appeal it under the same procedures which apply to appeals of the denial of records.

Your request for the waiver should give reasons why you should receive the waiver. Generally, you should argue that on receipt of the information you will disseminate it to the broader public and use it in a way that benefits them.

Getting State Information

As we discussed earlier, state agencies bear major responsibilities for permitting and other review of synfuel plant proposals under state law and under authority delegated by EPA pursuant to the Clean Air Act and Clean Water Act. As a result, state agencies will have important information about proposed plants.

The Freedom of Information Act does not apply to states, but 49 states have their own information laws (see next section). In addition, both the Clean Air and Clean Water Acts, and EPA regulations, require states exercising delegated permitting authority to make information available. The states have not consistently met this obligation.

If a state agency will not supply information, however, it may be possible to obtain it from EPA. Generally, states exercising delegated authority to administer air or water permits must supply permit application data to EPA. Once that information is in EPA's hands, you can use FOIA to get it. It does not matter that the information originally came from the state and that they do not want to give it to you. As long as the records are under the control of a federal agency, FOIA applies.

If you are trying to get your hands on such state records, file the request the moment the state records come into federal hands. If the Feds merely review them and then turn them back before you can get to them, they no longer have control. This may be one instance in which you want to be sticky about time deadlines. But if your request comes in at the right time, it is just like any other FOIA request and the fact that the information came from a state is immaterial.

State FOIA Laws

Forty-nine states have enacted freedom of information or open record laws. These laws vary widely in scope and procedure, although many are modeled on the federal statute..

If a permit is being administered by a state agency, or if some other important information about a project has been submitted to a state or other local unit of government (in a zoning or siting proceeding, for example), you may be able to obtain access to this information through the state law.

You will have to examine your state law to see if it is useful. Are the records available? Do any exemptions apply? What are the costs of copying or search for records? Some laws will obviously be more helpful than others in securing access.

State open record laws are listed below. Only Mississippi has no open record law. In every other state you are guaranteed access to at least some portion of state records.

State	Statutory Cite
Alabama	Title 41, Sections 145-147
Alaska	AS 09.25. 110 and 120
Arizona	Title 39, Chapter 1, Article 2
Arkansas	Sections 12-2801 to 12-1807
California	Government Code, Sections 6250-6260
Colorado	Title 24, Article 72, Sec. 203 et seq.
Connecticut	1-19, 1-20, Conn. Gen. Stat. Annotated
Delaware	Title 29, Sections 10001-10005
D.C.	D.C. Code 1-1501, et seq.
Florida	Chapter 119, F.S.A.
Georgia	Section 40-2701 et seq.
Hawaii	Hawaii Revised Statutes, Sections 92. 50-52
Idaho	Section 9-301, Idaho Code
Illinois	Chapter 116, Section 43
Indiana	Title 57-6, Indiana Code
Iowa	Chapter 68A, Code of Iowa
Kansas	Chapter 45, Section 201, Kan. Stat. Ann.
Kentucky	Chapter 61, Sections 870-884, Ken. Rev. Stat.
Louisiana	Title 44, Sections 31-38, Louisiana Code
Maine	Title I, Chap. 13, Sec. 405, Maine Rev. Stat.
Maryland	Art. 76A, Sections 1-5, Ann. Code of Maryland
Massachusetts	Title 10, Chap. 66, Massachusetts Code
Michigan	Section 750.492, Michigan Compiled Laws
Minnesota	Section 15.1621, Minn. Stat. Annotated
Mississippi	—
Missouri	Chapter 610, Sections 610.010-030
Montana	Title 93, Chapter 1001, Sections 93-1001-1-5
Nebraska	Section 84-712
Nevada	Sections 239.010-030
New Hampshire	Chapter 91-A

New Jersey	Section 47:1A
New Mexico	Sections 71-5-1—71-5-3, New Mexico Code
New York	Chapter 47, Section 85, et seq.
North Carolina	Gen. Stat., Chapter 132, Sec. 132-1-9
North Dakota	Section 44-04-18
Ohio	Section 149.34, Ohio Revised Code
Oklahoma	Title 51, Section 24
Oregon	Section 192.410-500
Pennsylvania	Sections 66.1-4
Rhode Island	38-2-1, et seq., R.I. Statutes
South Carolina	30-4-10, et seq., Code of Laws of S.C.
South Dakota	Title 1, Chapter 27, Sec. 1-7, S.D. Cod. Law
Tennessee	Sections 15-304-308, Tennessee Code
Texas	Art. 6252-17A, Sec. 1-13, Vernon's Tex. Civ.
Utah	Sections 78-26-1—8, Code of Utah
Vermont	1 V.S.A., Sections 315-320
Virginia	Sections 2.1-340—346.1, Code of Virginia
Washington	Title 42, Chap. 17, Sections 17.250 et seq.
West Virginia	Sections 29B-1-1—6
Wisconsin	Sections 19.21, 59.14, 59.71
Wyoming	16-4-201 et seq., Wyoming Statutes

FOIA and the Synthetic Fuels Corporation

The Energy Security Act, under which the U.S. Synthetic Fuels Corporation (SFC) was created, exempts the SFC from the Freedom of Information Act and substitutes a far weaker set of public disclosure standards in Section 121 of the Act.

PUBLIC ACCESS TO INFORMATION

Sec. 121. (a) The Corporation shall make available to the public, upon request, any information regarding its organization, procedures, requirements, and activities: Provided, That the Corporation is authorized to withhold information which is exempted from disclosure pursuant to subsection (b) of section 552 of title 5 United States Code, and section 116(f) of this part as it pertains to minutes of meetings of the Board of Directors.

(b) The Corporation, upon receipt of any request for information, shall determine promptly whether to comply with such request and shall promptly notify the person making the request of such determination. In the event of an adverse determination, and if requested by the person requesting the information, such determination shall be reviewed by the General Counsel of the Corporation.

(c) Section 1905 of title 18, United States Code, shall apply—

(1) to Directors, officers, and employees of the Corporation as if they were officers or employees of the United States; and

(2) to the Corporation as if it were a Federal agency.

The corporation has published guidelines on Disclosure and Confidentiality. The guidelines, on their face, appear not too dissimilar from those published by agencies covered by FOIA. But there are critical differences. All exemptions are mandatory. The corporation established a policy that allows submitters of information to make material confidential and have such classification honored until such time as there is a specific request for the material, which triggers a review. (This effectively keeps 90% of the application forms out of the public reading room where they belong.)

Time limitations for responding to requests and procedures for appeals (decided by the corporation's general counsel) are similar to those mandated by FOIA. But at that point the similarities end. The unhappy requester who has not received the material he asked for cannot go to Federal District Court and sue. The final SFC decision is essentially unreviewable.

If you are concerned about a synfuels project that has applied to the SFC for financial assistance, by all means ask for that information, but do not expect much.

THE ENERGY SECURITY ACT

The primary vehicle for federal subsidies to synthetic fuels projects is the United States Synthetic Fuels Corporation (SFC) established by the Energy Security Act of 1980, Public Law 96-294, 42 U.S.C. 8701 *et seq.* The authors of the Energy Security Act in the Congress and in the Carter Administration wanted massive federal assistance to synfuels made available fast. So they wrote the statute to minimize opportunities for citizens to influence or to sue the SFC. So, although it commits the taxpayers' money and exists by virtue of a federal law, in many ways the SFC is like a private bank.

How Much Money?

President Carter sought and Congress nominally approved an $88 billion synfuels program. But in reality Congress only made available $20 billion, of which approximately $15 billion remain to be committed by the SFC. That money can be committed by the SFC without Congressional or Presidential review.

The SFC's Legal Charter

The SFC is controlled by a seven-member, presidentially appointed board of directors. It has, as of July 1, 1982, a staff of 185. The chairman of the board serves as the chief executive officer.

Section 175(a) of the Energy Security Act states that:

> No federal law shall apply to the Corporation as if it were an agency or instrumentality of the United States, except as expressly provided in this part.

That means that laws requiring openness and fair procedures on the part of government agencies do not apply to the SFC, nor do most environmental laws (although most do apply to SFC projects to the same extent they would to any other).

The act also seeks to prevent suits by citizens to compel the SFC to comply with its own statute. That provision has not been tested in court, and may be vulnerable to challenge.

The SFC is authorized by the Energy Security Act to provide assistance to synthetic fuels projects in the form of

> (i) price guarantees, purchase agreements, or loan guarantees;
> (ii) loans; and
> (iii) joint ventures

in decreasing priority, §131(b) (2) (B). In selecting projects for support, the SFC must consider

> (A) the diversity of technologies (including differing processes, methods, and techniques); and
> (B) (i) the potential cost per barrel or unit production of synthetic fuel from the proposed synthetic fuel project;
> (ii) the overall production potential of the technology, considering the potential for replication, the extent of the resource and its geographic distribution, and the potential end use; and
> (iii) the potential of the technology for complying with applicable regulatory requirements.

The SFC has solicited and received proposals in two rounds, developed criteria for selecting among those proposals, and begun negotiations with two "finalists" among the first-round applicants. There has been little evidence thus far that the SFC will provide meaningful opportunities for citizens to influence the selection process, nor has the SFC evidenced a willingness to use its leverage to assure that the projects it selects will meet the highest possible environmental standards.

GLOSSARY

ACID GAS A gas which, when dissolved in ionizing liquid such as water, produces hydrogen ions. Carbon dioxide, hydrogen sulfide, sulfur dioxide, and various nitrogen oxides are the typical acid gases produced in coal gasification.

ACID GAS REMOVAL The process of selectively removing hydrogen sulfide and carbon dioxide from a gas stream.

ANTHRACITE COAL "Hard coal" containing a high percentage (86-98%) of fixed carbon and small percentages of volatile material and ash. It has a glossy black appearance and burns with little smoke. Nearly all anthracite in the U.S. is found in Eastern Pennsylvania.

AQUIFER An underground formation containing water.

AROMATIC HYDROCARBON A compound of carbon and hydrogen characterized by a ring of six carbon atoms, e.g., benzene.
Monocyclic aromatic hydrocarbons include benzene, toluene, and xylene (the so-called "BTX" family), of which benzene is a known human carcinogen. Hydrocarbons with a greater number of aromatic rings are called *polycyclic* aromatic hydrocarbons, or PAH, and are also linked to cancer. Notable among these are benzo[a]pyrene (BaP) and dibenz[a,h]anthracene, both animal carcinogens. Because of their larger ring structure, PAHs are sometimes called "heavier hydrocarbons."

ASH Theoretically, the inorganic salts contained in coal; practically, the noncombustible residue from the combustion of dried coal.

ASPHYXIANT A substance capable of producing a condition due to lack of oxygen in respired air, resulting in impending or actual cessation of life.

BENCH SCALE UNIT A small-scale laboratory unit for testing process concepts and operating parameters as a first step in the evaluation of a process.

BIOMASS An all-inclusive term, covering the many materials and substances, all having an origin in some living form, which constitute now, or could in the future, a source of energy. Included in this category are: wood and wood wastes, agricultural products and wastes such as stocks, stems, shells, cobs, or husks; algae, particularly large marine varieties such as kelp, animal wastes and municipal sewage.

BITUMEN Various mixtures of hydrocarbons (as tar) often occurring together with their nonmetallic derivatives. They either occur naturally or are obtained as residues after heat-refining naturally occurring substances, such as petroleum, asphalt or bituminous coal.

BITUMINOUS COAL The term is applied to coals of high and medium volatile matter content; the volatile matter varying from about 45 to about 25 percent. Fixed carbon contained ranges between 46 and 86%. The most plentiful, and comprises the majority of coal used in the United States. It ranges from dark grey to black in color, and produces smoke when burned. Bituminous coal is found in the Appalachian region, the Central Midwest, and some areas of the West.

BLOWDOWN Periodic or continuous removal of water containing suspended solids and dissolved matter from a boiler or cooling tower to prevent accumulation of solids.

BTU British Thermal Unit: The amount of heat needed to raise the temperature of one pound of water one degree Fahrenheit.

BTX Benzene, toluene, xylene; useful aromatic hydrocarbons produced and recovered in coal conversion processes.

CAKING The softening and agglomeration of coal as a result of the application of heat.

CARBON-HYDROGEN RATIO (C-H Ratio) The ratio, either on a weight or on a molecular basis, of carbon-to-hydrogen in a hydrocarbon material. Coal has a high carbon-hydrogen ratio, and coal conversion processes aim to lower that ratio, or increase the amount of hydrogen. A high carbon-hydrogen ratio means a solid material, like coal; lowering it converts the material to a liquid or a gas. The ratio is useful as a preliminary indication of the amount of hydrogen needed to convert the hydrocarbon to gases or liquids.

CARBONACEOUS MATERIAL A material rich in carbon.

CARBONIZATION Destructive heating of carbonaceous substances with the production of a solid, porous residue, or coke, and the evolution of a number of volatile products. For coal, there are two principal classes of carbonization, high-temperature coking (about 900°C) and low-temperature carbonization (about 700°C).

CATALYST Any substance, generally a solid, used to accelerate (usually) or retard a chemical reaction which does not itself undergo change during the process. "Catalyst poisoning" occurs when catalysts wear out or change too much.

CHAR The solid carbonaceous residue produced when coal is heated and/or reacted at temperatures sufficient to drive off its volatile matter. The major portion of the carbon does not volatilize or react and the remaining coke-like material is termed char.

CLEAN COLD GAS Producer gas that has been subjected to particulate, tar, and possibly sulphur removal processing steps.

COAL GAS The gas that comes from retorts, generators, gasifiers, or coke ovens during the distillation of coal. Large quantities of coal gas are produced as a byproduct when coal is used to make coke, coal tar, benzene, toluene, ammonia, and other products.

COAL LIQUIDS Liquids produced during the gasification, liquefaction or coking of coal. Coal liquids produced during coking are designated "by-product oils."

COAL TO METHANOL A generic term designating the conversion of coal through gasification with methanol as the end product. Catalytic conversion of the resulting syngas produces methanol.

COKE

Porous residue consisting of carbon and mineral ash formed when bituminous coal is heated in a limited air supply or in the absence of air. Coke may also be formed by thermal decomposition of petroleum residues. Coke is particularly useful in making iron and steel and as an industrial fuel.

COKE OVEN GAS

The gas secured from coke ovens during the production of coke. The properties of this gas are identical to those of low-BTU coal gas, and the two products are interchangeable.

COLD GAS EFFICIENCY

In gasification technology, this measurement is derived from the calorific value of the gas and by-products produced, divided by the sum of the energy inputs of raw materials, air, and steam.

COMMERCIALIZATION

The process during which a research and development project is converted into a self-sustaining, profitable business.

CRUDE OIL PROPERTIES

The chemical and physical properties of petroleum crude and synthetic oils. Principal properties include: API Gravity—A measure of the density of liquid petroleum products as defined by the American Petroleum Institute. The higher the API Gravity, the greater the number of gallons per pound of liquid. Viscosity—A measure of the internal friction or resistance of a liquid to flow. The higher the viscosity, the more difficult it is to pump and move the liquid.
Pour Point—The lowest temperature at which oil will pour.
C/H Ratio—The ratio of carbon atoms to hydrogen atoms in a hydrocarbon molecule.

DEMONSTRATION SCALE PLANT

A plant between pilot and commercial size built to demonstrate commercial feasibility of a process. Size is purely arbitrary. It is not uncommon, however, in conceptual designs of coal gasification plants, to use a single module of a larger commercial plant as the demonstration scale.

DESTRUCTIVE DISTILLATION

The distillation of coal accompanied by its thermal decomposition.

DISTILLATES

The liquid products condensed from vapor during distillation (as of coal liquids). Light distillates contain the lowest boiling constituents of the liquid, from which gasoline is produced. Middle distillates contain higher concentrations of the high boiling constituents, from which diesel and jet fuels are produced. Heavy distillates contain higher concentrations of the high boiling constituents, from which lubricating and residual oils are produced.

DISTILLATION A process of vaporizing a liquid and condensing the vapor by cooling used for separating liquids into various fractions according to their boiling points or boiling ranges.

EASTERN COAL Coals originating east of the Mississippi River. These coals are generally ranked higher in heating value, sulphur and caking tendencies than western coals.

ELECTROSTATIC PRECIPITATOR A device that uses an induced electrical charge to recover fine particles from a flowing gas stream.

ENTRAINED BED GASIFICATION A gasification process in which gasification of the candidate fuel takes place as it is carried along in a moving mix of fuel and the gasifying media. Characteristic of this form of gasification is a finely divided fuel, e.g., pulverized coal, to ease entrainment of the fuel particles.

FINES In general, the smallest particle of coal or mineral in any classification, process, or sample of material; especially those that are elutriated from the main body of material in the process.

FISCHER ASSAY The standard method of estimating oil content in shale-oil bearing rock. A representative sample of oil shale is crushed, dried, and heated according to specifications established by the U.S. Burea of Mines. The quantity of oil obtained is a measure of the "gallons-per-ton" yield or assay value. The actual amount of oil recovered from shale in any particular recovery process is expressed as a percentage of this Fischer assay value.

FISCHER-TROPSCH PROCESS A process in which synthesis (hydrogen and carbon monoxide in the ratio of about two mols of hydrogen to one mol of carbon monoxide at pressures on the order of 330 psig and temperatures of about 450°F) is used to produce synthetic liquids. The molecular weight of the product liquids (alcohols, fuel oils, gasoline feedstock, etc.) and, to a certain extent, the distribution of synthetic liquids and synthetic fuel gas can be shifted somewhat by appropriate adjustment of the reaction conditions.

FIXED BED GASIFICATION A gasification process in which the fuel is fed as sized lumps and in which the gas moves through a nearly stationary bed of reacting fuel.

FUEL GAS (Stack Gas) Synonymous terms for the gases resulting from combustion of a fuel.

FLUIDIZED BED GASIFICATION A gasification process in which fuel is suspended in a "boiling" bed by the gasifying media.

FUEL GAS Low heating value (150-350 Btu/scf) product generally utilized onsite for power generation or industrial use.

GAS LIQUOR (Sour Water) The aqueous acidic streams condensed from the coal conversion and processing areas by scrubbing and cooling of the crude gas stream.

GASIFICATION In fuel technology, the conversion of solid or liquid hydrocarbon to a fuel gas.

GASIFIER The processing unit in a synthetic fuels process in which the fuel, either solid or liquid, is converted to a synthetic gas (i.e., coal gasifier).

GRASS ROOTS PLANT A totally new facility built on what was an undeveloped site.

HEATING VALUE The energy released upon combustion (oxidation) of a fuel under standardized test conditions.

HIGH BTU GAS A term used to designate fuel gases having heating values of pipeline specification, i.e., greater than about 900 Btus per standard cubic foot. Natural gas has a heating value of about 1,050 Btu/scf.

HOT RAW GAS A manufactured gas that has not been cleaned or treated. It exists in the state that it leaves the gasifier.

HYDROCARBONS Any one of a very large class of chemical compounds composed of carbon and hydrogen. The largest single source of hydrocarbons today is petroleum crude oil. Coal and natural gas are also hydrocarbons.

HYDROGENATION The addition of hydrogen to any chemical molecule; reactions occur in the presence of a catalyst under high temperatures and pressures.

HYDROTREATING A process in which petroleum, petroleum products, or petroleum-like products are treated with hydrogen. Hydrotreatment involves passing the oil over a metallic catalyst in the presence of hydrogen; the sulphur and nitrogen are converted to hydrogen sulfide and ammonia while the hydrogen content of the oil is increased. It has been suggested that hydrotreating some synfuels may reduce their carcinogenicity or mutagenicity.

INDIRECT LIQUEFACTION — The generation of liquid hydrocarbon products from solid hydrocarbon fuels by first producing synthesis gas which is then further reacted to produce liquids.

IN SITU COAL GASIFICATION — Gasification of coal underground by introduction of air or oxygen into the coal seam.

KEROGEN — The organic material contained in oil shale; a hydrocarbon with significantly more hydrogen than coal.

LEACHING — The process of extracting a soluble component from a mixture by percolating a solvent, usually water, through the mixture, resulting in the solution and eventual separation of the soluble components.

LIGNITE — Brownish-black coal characterized by high moisture content and/or high volatiles—from about 30% to about 50%. Lignite may be defined as a solid fuel more mature than peat but less mature than bituminous coal. Other characteristics are high oxygen content and low content of fixed carbons. Maximum heating value for lignite is 8300 Btu on a moist and mineral matter free basis. Called "brown coal" in Germany.

LIQUEFACTION — The production of hydrocarbon liquids from solid fuels. Generally, conversion of a solid to a liquid; with coal, this appears to involve the thermal fracture of carbon-carbon and carbon-oxygen bonds, forming free radicals. These radicals abstract hydrogen atoms yielding low molecular weight gaseous and condensed aromatic liquids.

LIQUIFIED PETROLEUM GAS (LPG) — Those hydrocarbons that have a vapor pressure (at 70°) slightly above atmospheric (such as propane and butane); kept in a liquid form under pressure higher than 1 atmosphere.

LOCKHOPPER — A mechanical device that permits the introduction of a solid into an environment at different pressure.

LOW-BTU GAS — A term used to designate fuel gases having heating values from about 90 Btus per standard cubic foot to about 200 Btus per scf.

MAF — Moisture and ash free; a term that relates to the organic fraction in coal.

MEDIUM BTU GAS — A term used to designate fuel gases having heating values from about 250 to 400 Btus per standard cubic foot.

METHANATION — The catalytic combination of carbon monoxide and hydrogen to produce methane (CH_4) and water.

METHANE — CH_4, a colorless, odorless, tasteless gas, lighter than air; the chief component of natural gas.

METHANOL — A one-carbon atom alcohol (CH_3OH) formerly prepared by the destructive distillation of wood. Process have existed since about 1920 to convert coal to methanol, and today methanol can be converted to gasoline. Formerly called wood alcohol, or methyl alcohol.

MODULAR SYNFUELS PLANT — A plant comprised of parallel lines or "modules" having no common facilities with other parts of the plant, with the exception of utilities. A modular coal gasification plant feeds coal in at a number of similar units and exports a gas and by-products.

MOVING BED — A body of solids in which the particles or granules of a solid remain in mutual contact, but in which the entire bed moves in piston-like fashion with respect to the containing walls (in contrast with fixed bed).

NAPHTHA — A hydrocarbon distillate boiling in the range of about 140 to 420°F. Naphthas are subdivided according to actual distillation range into light, intermediate, heavy, and very heavy virgin naphtha.

NATURAL GAS — Naturally occurring gas extracted from sedimentary structures, consisting mainly of methane and having a higher heating value of approximately 1,050 Btus per standard cubic foot.

NON-CAKING COAL — A coal that does not tend to pass through a sticky phase (agglomerate) when heated.

OIL SHALE — A layered sedimentary rock which contains abundant quantities of kerogen. When heated to temperatures above 900°F, the kerogen in the rock decomposes, releasing shale oil.

ON-STREAM OPERATING TIME — The time during which the entire plant is actually working at present conditions, as opposed to the time in which it is shut down for repairs, starting up, etc.

OVERALL PLANT THERMAL EFFICIENCY — The fraction of total energy input to a plant which is exported as usable energy in the form of gas, by-products, and electricity.

PAH Polycyclic Aromic Hydrocarbon. See Aromatic Hydrocarbon.

PDU Process Development Unit; a system used to study the effects of process variables on performance; sized between a bench-scale unit and a pilot plant.

PEAT Peat is a marshy accumulation of partly decayed vegetable matter. It differs from coal in that it has not had the heat and pressure necessary to change its physical form to that of coal or lignite.

PETROCHEMICALS Those derived from crude oil or natural gas, or their coal-derived substitutes; they include light hydrocarbons such as butylene, ethylene, and propylene, the raw materials for the production of plastics by polymerization.

PHENOLS A group of aromatic compounds having the hydroxyl (OH) group directly attached to the benzene ring.

PILOT PLANT A small-scale industrial process unit operated to test the application of a chemical or other manufacturing process under conditions that will yield information useful in design and operation of full-scale manufacturing equipment.

POKEHOLE An opening in the cover of a process vessel through which steel rods are inserted, for the purpose of determining the fire bed depth and the ash bed depth in a gasifier.

POLYCYCLIC ORGANIC COMPOUND A compound whose molecular structure contains two or more rings (usually fused) that are mostly constructed of carbon atoms (e.g., anthracene). See also Aromatic Hydrocarbon.

PRESSURE GASIFICATION The conversion of a solid or liquid hydrocarbon to a gas at pressures greater than atmospheric.

PROCESS STREAM Any material stream within the coal conversion processing area.

PRODUCER GAS An industrial fuel made by processing air and steam continuously through the hot fuel bed of a gas producer. Consists essentially of carbon monoxide and hydrogen (50%) and nitrogen (50%).

PYROLYSIS Chemical change brought about by the action of heat in the absence of air.

QUAD A quadrillion Btus, the equivalent of one trillion cubic feet of natural gas. The U.S. uses about 75-80 quads of energy a year.

QUENCHING Cooling by immersion in oil, water bath, or water spray.

RETORT Any closed vessel or facility for heating a material for purposes of chemical reaction.

RETORTING Any of a variety of methods by which a carbonaceous material is heated, generally about 700°F, to effect decomposition of the material into gases, oils, tars, and carbon. Heat for retorting is obtained by combusting a portion of the raw feed and/or fuels derived from the process. Oil shale is processed by retorting.

RETROFITTING A term used in the combustion industry for adaptions and/or changes to existing equipment to permit substitution of one fuel for another.

RUBBLIZATION The process of breaking large solid materials into rough, broken pieces. As applied to oil shale technology, the process in which shale within a given volume is explosively fractured underground.

SASOL An acronym for South African Synthetic Oil Limited, an early name for what is today called the South African Coal, Oil, and Gas Corporation, Ltd.; state owned synthetic fuel facilities located in Sasolburg and Secunda, Republic of South Africa. South African coal is gasified in Lurgi gasifiers and a broad range of synthetic fuels and petrochemicals is produced.

SCRUBBER Apparatus in which a gas stream is freed of tar, ammonia, and hydrogen sulfide.

SENSIBLE HEAT The heat absorbed or released when a substance undergoes a change in temperature and does not undergo a phase change (i.e., liquid to gas, solid to liquid, etc.).

SHALE A fine-grained clay rock with slate-like cleavage, sometimes containing an organic oil-yielding substance called kerogen.

SHALE OIL Oil derived by the pyrolysis of the organic matter in oil shale.

SHIFT CONVERSION Process for the production of gas with a desired carbon monoxide content from crude gases derived from coal gasification; carbon monoxide-rich gas is saturated with steam and passed through a catalytic reactor where the carbon monoxide reacts with steam to produce hydrogen and carbon dioxide, the latter being subsequently removed in a scrubber employing a suitable sorbent.

SLAG Molten coal ash composed primarily of silica, alumina, iron oxides, and calcium and magnesium oxides.

SLURRY A suspension of pulverized solid in a liquid.

SLUDGE A soft mud, slush, or mire, e.g., the solid product of a filtration process before drying.

SOLVENT EXTRACTION A means of separating mixed materials by dissolving one of them in a solvent.

SOUR GAS A gas containing acidic substances such as hydrogen sulfide or carbon dioxide.

SPENT SHALE The retorted residual material after the oil and gas products are removed. Its properties vary with the type of retorting procedure used; indirectly heated retorts produce a carbonaceous spent shale, while directly heated retorts produce a material essentially stripped of carbon.

SCF Standard cubic foot; the volume of a gas at standard conditions of temperature and pressure.

SUB-BITUMINOUS COAL Coal of intermediate rank (between lignite and bituminous); weathering and nonagglomerating coal having calorific values in the range of 8,300-11,000 Btu, calculated on a moist, mineral/matter-free basis.

SURFACE RETORTING In the oil shale industry, the processing of shale in above ground retort vessels.

SWEET GAS Gas from which acidic constituents such as H_2S have been removed.

SYNCRUDE A crude oil derived from processing a carbonaceous material. For example, oil extracted from shale and unrefined oil from coal conversion plants are termed syncrudes.

SYNTHESIS GAS or SYNGAS — A mixture of carbon monoxide and hydrogen which can be reacted to yield hydrocarbons used in chemical synthesis.

SYNTHETIC FUELS — Fuels which exist in physical and chemical forms different from those in the naturally occurring state. Strictly speaking, modern automobile gasolines and diesel fuels would fall in this category. Any fuel made from coal, unconventional oil resources or fermented farm products.

SYNTHETIC NATURAL GAS (SNG) — Substitute for natural gas (which is mostly methane); a manufactured gaseous fuel generally produced from naphtha or coal that contains 95% to 98% methane and has an energy content of 980 to 1,035 Btu/scf (about the same as that of natural gas). Because of its high quality, and thus its "methane interchangeability," SNG can be transported long distances in natural gas pipelines.

TAIL GAS — A gas issuing from a gas-treatment unit which may be recycled to the process or exhausted.

TAR — A brown or black, viscous, combustible liquor formed by the destructive distillation of coal. It condenses out of the raw gas stream as part of the gas liquor, has a specific gravity of approximately 1.1, and contains most of the fines which are carried over from the gasifier in the gas stream.

TAR OIL — The more volatile portion of the tar, with a specific gravity of approximately 0.9, a boiling range of approximately 185-300°C (365-660°F) depending on the coal feed and operation conditions. In addition, tar oil floats on the gas liquor.

TAR SANDS — Large deposits of finely divided clay and/or siliceous materials coated with layers of bitumen and water. They are tar-like in appearance.

TRACE ELEMENTS — This term is applied to elements that are present in the earth's crust in concentrations of 0.1% (1000 ppm) or less. Trace element concentrations are usually somewhat enriched in coal ash. Enviromentally hazardous trace elements present in coal include: antimony, arsenic, beryllium, cadmium, mercury, lead, selenium, and zinc.

VENTING — Release of gases or vapors under pressure to the atmosphere.

VENTURI SCRUBBER A gas cleaning device which involves the injection of water into a stream of dust-laden gas flowing at a high velocity through a contracted portion of a duct, thus transferring the dust particles to the water droplets which are subsequently removed.

WATER GAS Gas produced by the reaction of carbon (in coal or coke) and steam to yield mixtures of carbon monoxide and hydrogen; similar to synthesis gas.

WATER GAS SHIFT The reaction between water vapor and carbon monoxide to produce hydrogen and carbon dioxide or the reverse: $CO + H_2O = H_2 + CO_2$.

Sources:

Dravo Engineers and Constructors, *Synfuels Glossary*, Pittsburgh, PA, April 1981. Used by permission.

U.S. Department of Energy, *Coal Liquefaction Quarterly Report*, Washington, D.C., May 1979.

National Institute of Occupational Safety and Health, *Criteria for a Recommended Standard—Occupational Exposures in Coal Gasification Plants*, DHEW (NIOSH) 78-191, Cincinnati, OH, September 1978.

U.S. Congress, Office of Technology Assessment, *An Assessment of Oil Shale Technologies*, two volumes, Washington, D.C. June 1980.

Notes

2. History

Oil Shale

1. "West Germany's BKB May Build Oil Shale Pilot Plant in Schandelah," *Synfuels,* 3 July 1981, 3-4.

2. U.S. Congress, Office of Technology Assessment, *An Assessment of Oil Shale Technologies,* 2 volumes, Edward M. Perrini, *Oil from Shale and Tar Sands* (Park Ridge, N.J., Noyes Data Corp., 1975), [hereafter OTA] 1; Anthony Sampson, *The Seven Sisters,* revised edition (New York:Bantam, 1979), 134.

3. OTA, I-108.

4. "Sweden to Launch Oil Shale Pilot in 1982; Commercial Plants by 1990s," *Synfuels,* 10 July 1981, 8-9.

5. "West Germany's BKB May Build Oil Shale Pilot Plant in Schandelah," *Synfuels,* 3 July 1981, 4.

6. Powder River Basin Resource Council, *Target: Wyoming: A Study of Synthetic Fuel Development in Wyoming,* Sheridan, WY, 1981, 3.

7. Marc Reisner, "The Rock That Burns, The Coal That Flows," *The Amicus Journal* 2:3, Winter 1981, 16.

8. Thomas N. Bethell, "How to Keep it Going: Synfuels," *Washington Monthly* 12:8, October 1980, 18.

9. R.F. Cane, "The Origin and Formation of Oil Shale," in T.F.Yen and G.C. Chilingarian, *Oil Shale* (Amsterdam: Elsevier Scientific Publishing Co., 1976, 56-57.
10. *Ibid.*
11. OTA I-108-110.
12. OTA, I-108-110; see also John D. Baker, [hereafter Baker], "World Oil Shale Resources and Development History," *Symposium Papers: Synthetic Fuels from Oil Shale,* Institute of Gas Technology, April 1980, 9.
13. OTA, I-108. 14. OTA, I-108, 109, 139; Baker, 5. 15. OTA, I-112. See also "Brazilian Shale Prototype," *Synfuels,* 3 April 1981, 1.
16. *Ibid.*
17. *Ibid.*
18. Baker, 10.
19. OTA, I-111.
20. See "Estonia Testing 3000-Tons/Day," *Synfuels,* 3 July 1981, 9-10; Britt J. Vesilind, "Return to Estonia," *National Geographic,* April 1980, 484-511; "Soviets Admit Troubles in Oil Shale," *The Oil and Gas Journal,* 6 October 1975, 42-3.
21. OTA, I-110.
22. *Ibid.*
23. Not all claims are perfected and patented, but by 1927 more than 175,000 acres of land had been transferred. Many of the unperfected claims are still potentially valid according to a recent U.S. Supreme Court ruling, *Andrus* v. *Shell,* (1980).
24. Robin Bates, producer, "Do We Really Need the Rockies?" transcript of PBS-TV's "Nova" program, broadcast 28 October, 3. [hereafter Nova], 3.
25. NOVA, 3-4.
26. OTA, I-110; U.S. Federal Energy Administration, Project Independence, Interagency Task Force on Synthetic Fuels From Coal, *Final Task Force Report*, Washington, D.C., November 1974, xiii.
27. OTA, I-111.
28. NOVA, 21.
29. OTA, I-139-140; Office of Technology Assessment, *Draft Working Paper for a Case Study of Oil Shale Technology: Oil Shale Retorting Technology, Washington, D.C., March 1980,* [hereafter: OTA Draft Working Paper], 31; EPA Oil Shale Research Group, *Environmental Perspective on the Emerging Oil Shale Industry, EPA 6002-80-205a, Cincinnati, Ohio, 1980, 17.*

Gas from Coal

1. U.S. Federal Energy Administration, Project Independence, Interagency Task Force on Synthetic Fuels from Coal, *Final Task Force Report,* Washington, D.C., November 1974, xiii.
2. Ogden Hammond and Robert F. Baron, "Synthetic Fuels: Prices, Prospects and Prior Art," *American Scientist* 64:4, July-August 1976 [hereafter "Hammond and Baron"], 407.
3. Malcolm W. H. Peebles, *The Evolution of the Gas Industry* (Hong Kong: New York University Press, 1980) [hereafter "Peebles"], 7.

4. Devra Lee Davis, "Cancer in the Workplace: The Case for Prevention," *Environment* 23:6, July/August 1981, 26.

5. Colin A. Russell, *Coal, the Basis of Nineteenth Century Technology* (Block II, Unit 4 of "Science and the Rise of Technology Since 1800) Bletchley, England: The Open University Press, 1973 [hereafter "Russell"], 38.

6. Peebles, 8.

7. *Ibid.*

8. William A. Bone and Godfrey W. Himus, *Coal: Its Constitution and Uses* (London: Longmans, Green & Co., 1936) [hereafter "Bone"], 327.

9. Bone, 327-8.

10. Bone, 328.

11. Russell, 25.

12. *Ibid.*

13. Michael Barash and Walter J. Gooderham, *Gas* (Exeter: Wheaton-Pergamon, 1971 [hereafter "Barash"], 10.

14. Russell, 25.

15. Bone, 328.

16. Peebles, 10; Bone, 329.

17. Bone, 329.

18. Rusell, 27.

19. Russell, 35.

20. Hammond and Baron, 407.

21. Russell, 28.

22. *Ibid.*

23. Peebles, 16-17.

24. Barash, 6.

25. Russell, 27-8.

26. Richard Doll, "The Causes of Death Among Gas-Workers with Special Reference to Cancer of the Lung," *British Journal of Industrial Medicine* 9:3, July 1952, [hereafter "Doll 1952"], 180.

28. Doll 1952, 183.

30. U.S. Department of Health, Education and Welfare, National Institute of Occupational Safety and Health, *Criteria for a Recommended Standard . . . Occupational Exposure to Coal Tar Products*, DHEW-NIOSH Pub. No. 78-107, Rockville, MD, September 1977, [hereafter "Coal Tar"], 28.

31. Richard Doll, et al., "Mortality of gasworkers wih special reference to cancers of the lung and bladder, chronic bronchitis, and pneumoconiosis," *British Journal of Industrial Medicine* 22:1, January 1965 [hereafter "Doll 1965"], 2.

32. Doll 1965, 3-4.

33. Doll 1965, 4.

34. Coal Tar, 67.

35. Richard Doll, et al., "Mortality of gasworkers—final report of a prospective study," *British Journal of Industrial Medicine* 29:4, October 1972 [hereafter "Doll 1972"], 396.

36. Doll 1972, 403.

37. Doll 1972, 404.
38. Doll 1972, 405.
39. *Ibid.*
40. Douglas Costle, "EPA and Synthetic Fuels," presentation to Board of Directors of the U.S. Synthetic Fuels Corporation, 22 December 1980, Washington, D.C.
41. "Fear of the dole keeps Britain's dirtiest factory open," *New Scientist* 90:1258, 18 June 1981, 745.
42. Russell, 35.
43. Russell, 38.
44. Russell, 35.
45. Peebles, 8.
46. Peebles, 10.
47. Peebles, 53.
48. Howard N. Eavenson, *The First Century and a Quarter of American Coal Industry* (privately printed, Koppers Building, Pittsburgh, PA, 1942) [hereafter "Eavenson"], 188.
49. Peebles, 53-54.
50. Eavenson, 202, 297, 334, 357.
51. Anthony Sampson, *The Seven Sisters,* revised edition (New York: Bantam, 1979), 45.
52. Hammond and Baron, 415.
53. Peebles, 54.
54. Robert R. Hershman, "Chronology of Gas Company Events," Washington Gas Light Company, Washington, D.C., 9 Sept. 1946, 54.
55. James Hardcastle, "Priming the Pump," *Science 81,* January-February 1981, 58-63.
56. Enviro Control, Inc., *Coal Gasification and Industrial Hygiene Characteristics: A Literature Review to-March 1979,* unpublished report submitted in draft form to NIOSH, Rockville, MD, March 1979 [hereinafter "ECI Gas"], I-1.
57. U.S. Department of Energy, "Gas From Coal," Fossil Energy Fact Sheet, DOE/FE-007, Washington, D.C., June 1980 [hereafter "Gas Facts"], 1.
58. William F. Hederman, Jr., *Prospects for the Commercialization of High-Btu Coal Gasification,* prepared for DOE, R-2294-DOE, Rand Corporation, Santa Monica, CA, April 1978, 1.
59. Gas Facts, 1.
60. U.S. Congress, Congressional Research Service, "Synfuels from Coal and the National Synfuels Production Program: Technical, Enviromental, and Economic Aspects," prepared for the Senate Committee on Energy and Natural Resources, Pub. No. 97-3, Washington, D.C., January 1981 [hereafter "CRS"], 24.
61. Joanne Omang, "Early Plant's Waste Floats in Trout Stream," *Washington Post,* 28 May 1981, A2.
62. Masatake Kawai, et al., "Epidemiologic Study of Occupational Lung Cancer," *Archives of Environmental Health* 14:6, June 1967 [hereafter "Kawai"], 859.
63. Frederica Perera, "Benzo(a) pyrene," internal NRDC memorandum, 13 May 1980, A-10.
64. Kawai, 862.
65. Doll 1965, 8.

66. Gas Facts, 1.

67. Kennedy Maize, "Exxon Bucks The Industry's Gloom Over Synfuels Future," and "It's Lurgi for Exxon Lignite," *Energy Daily*, 7 August 1981, 3.

68. CRS, 24.

69. Peebles, 13.

70. Harry Perry, "The Gasification of Coal," *Scientific American* 230:3, March 1974 [hereafter "Perry"], 19-20.

71. Peebles, 15; see also Perry, 20.

72. ECI Gas, I-3.

73. James S. McIlhenny, "History of the Washington Gas Light Company," interdepartmental memorandum to D. D. Ransdell, Washington Gas Light Company, Washington, D.C., 16 February 1932, 1; Peoples Gas Light and Coke Company, *Gas News*, Chicago, IL, September 1950, 1.

74. ECI Gas, I-3.

75. Sources for the brief account of modern gasifiers include Alan Brainard, ed., *Proceedings: Seventh Annual International Conference on Coal Gasification, Liquefaction and Conversion to Electricity, August 5-7, 1980,* University of Pittsburgh, Pittsburgh, PA, 1980; Tennessee Valley Authority, *Symposium: Ammonia from Coal, May 8-10, 1979,* Bulletin Y-143, National Fertilizer Development Center, Mussel Shoals, AL, July 1979; *ECI Gas,* I-3 to I-9, I-13 to I-16; M. Ghasshemi, et al., TRW Inc. *Environmental Assessment Data Base Per High-Btu Gasification Technology,* three volumes, prepared for EPA, EPA-600/7-78-186a-c, Cincinnati, OH, September 1978, A-3, A-179-181; C.A. Zee, et al., "Environmental Assessment: Source Test and Evaluation Report & Koppers-Totzek Process" (Modderfontein, RSA), with an appendix from Krupp-Koppers GMGH, prepared for EPA, EPA-600/7-81-009, Cincinnati, OH, January 1981; "Private Negotiations Break Down in Koppers-Totzek Suit," *Syn Fuels,* 10 October 1980, 1; "Lurgi Leads Market, K-T Follows," *Synfuels Week*, 17 August 1981, 2; United Nations Economic Commission for Europe, Coal Committee, "Working Meeting II: Coal/Sem. 3/R.7, 17 October 1975, Dusseldorf, in *Symposium on Gasification and Liquefaction of Coal* (Essen, FR6: Verlag Gluckauf GMGH, 1976).

76. ECI Gas, I-3.

77. ECI Gas, I-6.

78. ECI Gas, I-7.

79. ECI Gas, I-9, I-16.

80. Perry, 20.

81. ECI Gas, I-27.

82. C. A. Zee, et al., TRW, Inc., "Environmental Assessment: Source Test and Evaluation Report, Koppers-Totzek Process" (Modderfontein, RSA), prepared for EPA/ERL, EPA 600-7-81-009, January 1981, 2.

83. "Lurgi leads market, K-T follows," *Synfuels Week,* 17 August 1981, 2.

84. Enviro Control, Inc., "Tab J: Trip Reports," submitted to NIOSH, December 1977, 65-66.

85. National Institute of Occupational Safety and Health, *Criteria for a Recommended Standard . . . Occupational Exposures in Coal Gasification Plants,* DHEW (NIOSH) 78-191, Rockville, MD, September 1978, 44.

86. *Ibid.*

Liquids from Coal

1. Lawrence Hoffman, et al., *Environmental, Operational and Economic Aspects of Thirteen Selected Energy Technologies,* prepared for EPA, EPA-6001 7-80-173, Washington, DC, September 1980, 105.

2. Ogden Hammond and Robert F. Baron, "Synthetic Fuels: Prices, Prospects and Prior Art," *American Scientist* 64:4, July-August 1976 [hereafter "Hammond and Baron"], 407.

3. U.S. Department of Energy, "Solvent Refined Coal-II: Turning Coal into Clean Energy," Gulf Oil Corporation, Pittsburg, PA, undated, 1.

4. *Ibid.*

5. Hammond and Baron, 407.

6. U.S. Department of Energy, "SRC-II Background and Schedule" Washington, DC, undated.

8. Hammond and Baron, 407, 415.

9. Enviro Control, Inc., *Coal Liquefaction and Industrial Hygiene Characterization: A Literature Review to October 1978*, unpublished report submitted in draft form to NIOSH, Rockville, MD, March 1979 [hereafter "ECI Liquids"], I:2-3.

10. ECI Liquids, I-3.

11. Enviro Control, Inc., *Coal Gasification and Industrial Hygiene Characteristics: A Literature Review to March 1979*, unpublished report submitted in draft form to NIOSH Rockville, MD, March 1979 [hereafter "ECI Gas"], I:2.

12. Joseph Borkin, *The Crime and Punishment of I. G. Farben,* New York: Free Press, 1978 [hereafter "Borkin"], 4.

13. Borkin, 9.

14. Borkin 17.

15. Borkin, 19-20.

16. Borkin, 21.

17. ECI Liquids, 1:4.

18. ECI Liquids, I:5.

19. Borkin, 26.

20. Borkin, 45.

21. Borkin, 45.

23. Borkin, 43.

24. Borkin, 45.

25. Borkin, 46.

26. *Ibid.*

27. Borkin, 49-50.

28. Borkin, 53.

29. Borkin, 55.

30. Borkin, 56-57.

31. Borkin, 60.

32. Borkin, 69.

33. David Masselli, "Synfuels Madness Makes a Comeback," and "Synthetic Fuels and the German War Effort," *Not Man Apart,* August 1979 [hereafter "Masselli"], 3.

34. ECI Liquids, 1:4.

35. Johannes Meintjes, *Sasol: 1950-1975* (Cape Town, RSA: Tafelberg-Uitgewers Beperk, 1973 [hereafter "Meintjes"], 153-154.

36. Masselli, 3.

37. ECI Liquids, 1:3.

38. *Ibid.*

39. ECI Liquids, I-6; Anthony Sampson, *The Seven Sisters*, revised edition (New York:Bantam, 1979), 49; W.R.K. Wu and H.H. Storch, *Hydrogenation of Coal and Tar*, U.S. Dept. of Interior, Bureau of Mines Bulletin No. 633, Washington, DC, 1968, 3-4; William A. Bone and Godfrey W. Himus, *Coal: Its Constitution and Uses* (London: Longmans, Green, & Co., 1936) , 552-554; Borkin, 89-94; G.S. Gibbs and E.H. Knowlton, *The Resurgent Years: 1911-1927* (New York: Harper Bros., 1956), 544-546.

40. Meintjes, 156, 157.

41. Meintjes, 153.

42. Meintjes, 13.

43. Masselli, 3.

45. ECI Liquids, I:7.

46. ECI Liquids, 1:6.

47. Borkin, 64, 68.

48. Masselli, 3.

49. Borkin, 53,60.

50. ECI Liquids, B:26.

51. Christopher Joyce,"The hazards of synthetic fuels," *New Scientist*, 85:1192, 31 January 1980, 300.

52. ECI Liquids, I:10.

53. R. H. Weil and J. C. Lane, *Synthetic Petroleum from the Synthane Process* (Brooklyn:Chemical Publishing Co., 1948), 204.

54. ECI Liquids, 1:10.

55. U.S. Department of Interior, Bureau of Mines, *Synthetic Liquid Fuels, Annual Report of the Secretary of the Interior for 1952, Part I: Oil from Coal*, Bureau of Mines Report of Investigations 4942, Washington, DC, January 1953 [hereafter "Bureau of Mines 4942"], vii-viii. 56. Bureau of Mines 4942, 4, 5, 8, 11, 12.1

57. Bureau of Mines 4942, 17-18, 26.

58. Bureau of Mines 4942, 23.

59. U.S. Department of the Interior, Bureau of Mines, *Synthetic Liquid Fuels, Annual Report of the Secretary of the Interior for 1953, Part II: Oil Shale*, Bureau of Mines Report of Investigations 5044, Washington, DC, April 1954 [hereinafter "Bureau of Mines 5044"], ii. 60. *Ibid.*

61. Richard J. Sexton, "The Hazards to Health in the Hydrogenation of Coal," *Archives of Environmental Health* 1:3, September 1960, 1:186.

62. ECI Liquids, 1:42.

63. Sexton IV, 224.

64. Alan Palmer, "Mortality Experience of 50 Workers with Occupational Exposures to the Products of Coal Hydrogenation Processes," *Journal of Occupational Medicine*, vol. 21, 1979, 41-44.

65. ECI Liquids, 1:15.

66. ECI Liquids, 1:13.
67. ECI Liquids, 1:16.
68. Meintjes, 12, 13.
69. Meintjes, 20.
70. Meintjes, 33-34.
71. Meintjes, 66.
72. Meintjes, 42-45.
73. Richard Myers, "Synthetic Fuels—Direct, Indirect Liquefaction: The Debate Marches On," *Energy Daily*, 20 August 1979 [hereafter "Myers"], 4.
74. Meintjes, 74.
75. Meintjes, 154-55.
76. Myers, 4.
77. Meintjes, 159.
78. Meintjes, 88, 96.
79. Meintjes, 97.
80. Meintjes, 137.
81. Meintjes, 136-138
82. Meintjes, 137.
83. Meintjes, 141.
84. *Ibid.*
85. Myers, 4.
86. Enviro Control, Inc., *Trip Report, Sasol I, Sasolburg, South Africa,* 5-7 December 1977, submitted to NIOSH, Rockville, MD, 1978 [hereafter "Sasol trip"], 16, 23.
87. Sasol trip, 22, 23.
88. Sasol trip, 21.

3. Technology of Synthetic Fuels Production

1. For example, high Btu coal gasification plants have a thermal efficiency of only 45 to 65%, depending on the process. This means that the plant consumes about one unit of fuel to produce a unit. By comparison, low Btu gasifiers are 85 to 90%, and direct liquefaction plants are between 60 and 75% efficient.

2. These are rough numbers. The actual amount of fuel produced per ton of coal or shale varies widely depending on the characteristics of the coal or shale, the products being produced by the synfuel plant, and the technology used.

3. Detailed but readable analyses of some of the more important of these technologies can be found in David Masselli and Norman L. Dean, Jr., *The Impacts of Synthetic Fuels Development*, Washington, DC, National Wildlife Federation, 1981, Appendix A; The Aerospace Corp., *Energy Technologies and the Environment*, prepared for DOE, DOE/EP-0026, Washington, DC, June 1981; [hereafter "Aerospace"] Congressional Research Service, *Synthetic Fuels from Coal: Status and Outlook of Coal Gasification and Liquefaction*, Washington, DC, 1979, I-1 to II-30; *New Technologies for Old Fuels*, Subcommittee on Fossil and Nuclear Energy Research, Development, and Demonstration of the House Committee on Science and Technology, Serial No. 25, 95th Cong., 2nd Session, 1977; Edward J. Bentz and Eliahu J. Salmon, *Synthetic Fuels Technology Overview* [hereafter "Overview"], Ann Arbor, MI, Ann Arbor Science, 1981, 13-53.

4. It is also theoretically possible to provide the heat needed in the chemical reaction by some other method such as the use of electrically heated elements in the gasifier or the injection of coal into molten salts or iron. However, in most of the gasifiers developed to date, the needed heat is provided by burning part of the coal in the gasifier.

5. A number of important chemical reactions take place in gasifiers. Carbon and oxygen combine to produce carbon monoxide and carbon dioxide. Carbon monoxide and hydrogen are created through two different reactions: the reaction of steam and carbon, and the reaction of water and carbon monoxide. Methane is produced as the result of the reaction of carbon with hydrogen and the reaction of carbon monoxide with hydrogen.

6. In some medium Btu gasification processes, hot hydrogen is injected into the gasifier instead of steam. This *hydrogasification* process produces more methane than simple gasification and consequently the resulting gas requires less processing to be converted into high-Btu gas.

7. Catalysts are commonly employed in synfuel technologies. A catalyst is a chemical substance that accelerates the rate of a chemical reaction without itself undergoing any permanent chemical change.

8. More specifically, in the shift conversion step, the carbon monoxide and water in the gas are reacted to produce hydrogen, carbon dioxide, and heat. Enough of the synthesis gas is shifted in this fashion to set up a ratio of three molecules of hydrogen to one molecule of carbon monoxide.

9. Most high-sulfur bituminous coals found in the eastern United States are caking coals. Western sub-bituminous coals and lignites and European coals tend to be non-caking.

10. The following discussion is largely based on: National Institute for Occupational Safety and Health, *Criteria for a Recommended Standard . . . Occupational Exposures in Coal Gasification Plants*, DHEW (NIOSH) 78-191, Cincinnati, OH, September 1978, 19-33 and 111-76.

11. Overview, 22.

12. *Hydrotreating* is a process of stabilizing and/or removing sulfur and other objectionable elements from a substance by reacting it with hydrogen in the presence of a catalyst.

13. Edward W. Merrow, et al., *Understanding Cost Growth and Performance Shoftfalls in Pioneer Process Plants (Santa Monica, CA: Rand Corporation, 1981)* and E. Hodson Thornber, Synthetic Fuel Plants: Risk Assessment and Financial Assistance Under the U.S. Synthetic Fuels Act of 1980 (unpublished, 1981).

14. U.S. Synthetic Fuels Corporation, *Staff Briefing for the Board on Strategic Matters,* unpublished, 8 January 1982.
15. *Ibid.*, 7.
16. Colorado Energy Research Institute and Colorado School of Mines Research Institute, *Oil Shale 1982: A Technology and Policy Primer* [hereafter "Oil Shale 1982"], Golden, CO, November 1981, 25. Much of this section was derived from this work.
17. For comprehensive yet readable summaries of oil shale technologies, see: Office of Technology Assessment, *An Assessment of Oil Shale Technologies,*[hereafter "OTA"] Washington, DC, June 1980, I: 119-176; C.C. Shih, et al., TRW Inc., *Technological Overview Reports for Eight Oil Shale Recovery Processes,* [hereafter "Shih"], prepared for EPA, Washington, DC, March 1979; and EPA Oil Shale Research Group, *Environmental Perspective on the Emerging Oil Shale Industry,* EPA-600/2-80-205a, Cincinnati, OH, 1980, 277-324.
18. Oil Shale 1982, 26.
19. Oil Shale 1982, 30.
20. Oil Shale 1982, 30-31.
21. Aerospace, 153.
22. In alternative true in situ processes, hot gases such as methane or steam are injected into the formation in order to "cook" the kerogen.
23. There is continuing research on various more exotic methods for cooking shale in place. For example, some companies are looking into the use of micro-waves, radio waves, bacteria, or special solvents for breaking down kerogen. However, none of these approaches is yet close to commercial development.
24. OTA, I-20.
25. Oil Shale 1982, 40.
26. OTA, I-120.

4. Environmental Impacts of Coal Conversion

1. Congressional Research Service, *Synfuels from Coal and the National Synfuels Production Program: Technical, Environmental, and Economic Aspects* [hereafter "National Program"], Publication No. 97-3, January 1981, 138.
2. James Antizzo, ed., Background Material for the Workshop on *Health and Environmental Effects of Coal Gasification and Liguefaction Technologies* [hereafter "Mitre"], Mclean, VA., The Mitre Corporation, 1978, 64.
3. *Mitre,* 69.
4. *Mitre,* 62.
5. U.S. Environmental Protection Agency, *Pollution Control Guidance Document for Lurgi-Based Indirect Liguefaction Facilities*—Staff Draft, vol. I [hereafter "Lurgi PCGD"] Research Triangle Park, N.C.: U.S. EPA, 15 May, 1981, xxiv-xxix.

6. U.S. Department of Energy, *Final Environmental Impact Statement Solvent Refined Coal-II Demonstration Project,* two volumes [hereafter "SRC-II"], Washington, D.C.: U.S. DOE, DOE/EIS-0069/V2, January 1981, C-58.

7. U.S. Department of Energy, *Final Environmental Impact Statement Solvent Refined Coal-I Demonstration Project,* two volumes, [hereafter "SRC-I"], Washington, D.C.: U.S. DOE, DOE/EIS-0073/V2, July 1981, C-66, C-33.

8. *Mitre,* 51.

9. U.S. Department of Energy, *Synthetic Fuels and the Environment: An Environmental and Regulatory Impacts Analysis* [hereafter "Synfuels"], Washington, D.C.: U.S. DOE, DOE/EV-0087, June 1980, 3-9.

10. W.H. Chesner, *et al., Assessment of the Potential Risk to Public Health, Safety and the Environment from Synfuel Solid Waste,* prepared for the Ohio River Basin Commission [hereafter "Chesner"], Cincinatti, Ohio, December 1980, D-2.

11. *Mitre,* 63.

12. Chesner, 3-16, 17.

13. U.S. Environmental Protection Agency, *Chemical and Biological Characterization of Leachates from Coal Solid Wastes,* Research Triangle Park, N.C.: U.S. EPA, EPA-600/7-80-039, March 1980, 79.

14. *Ibid.,* 4.

15. U.S. Environmental Protection Agency, *Symposium Proceedings: Environmental Aspects of Fuel Conversion Technology, III,* B. I. Loran and J.B. O'Hara, Ralph M. Parsons Co., "Specific Environmental Aspects of Fischer-Tropsch Coal Conversion Technology," in EPA-600/7-78-063, Research Triangle Park, N.C., April 1978, 420.

15A. *Lurgi PCGD,* 2-131.

16. U.S. Environmental Protection Agency, *Environmental Assessment Data Base for High-Btu Gasification Technology: Volume I* [hereafter "Gasification Data Base"05 #, Research Triangle Park, N.C.: U.S. EPA, EPA-600/7-78-186a, 1978, 96.

17. Ronald F. Probstein and Harris Gold, *Water in Synthetic Fuel Production: The Technology and Alternatives* [hereafter "Water"], (MIT Press, Cambridge, 1978, 94.

18. *Ibid.*

19. *Gasification Data Base,* 101.

20. *Water,* 9.

21. *Ibid.,* 132, 140, 181.

22. *SRC-II,* 4-35.

23. *Ibid.*

24. *Ibid.,* 4-37.

25. *Ibid.,* 4-44.

26. *Ibid.*

27. U.C. Berkeley Department of Chemical Engineering, *Processing Needs and Methodology for Wastewaters from the Conversion of Coal, Oil Shale and Biomass to Synfuels,* prepared for DOE, DOE/EV-0081, May 1980, 5-6.

28. *Synfuels,* 5-56.

29. *SRC-II,* 4-39.

30. U.S. Department of Energy, *Energy Technologies and the Environment—Environmental Information Handbook,* [hereafter "Handbook"], Prepublication Draft, DOE/EV/74010-1, Washington, D.C., December 1980, 7-17.

31. *SRC-II,* C-63.
32. *SRC-II,* C-62.
33. *SRC-II,* Z-40.
34. Battelle-Pacific Northwest Laboratory, *Ecological Fate and Effects of Solvent Refined Coal Materials: A Status Report* (Richland, WA.: Battelle-Pacific Northwest Laboratory PNL-3819), 144-148.
35. R.A. Pelroy, D.S. Sklarew, and S.P. Downey, *Comparison of the Mutagenicities of Fossil Fuels* (Richland, WA.: Batelle-Pacific Northwest Laboratory PNL-SA-9309), 10.
36. "Poisonous Chemicals Spilled at SRC-2 Pilot Plant," *The Energy Daily* (July 29, 1980), 1-2; Memorandum from Jim Oberlander and Jim Krull, Washington Department of Ecology, to Bruce Cameron and Lloyd Taylor, March 20, 1980.
37. *SRC-II,* Z-1.
38. *SRC-II,* Z-23.
39. *SRC-II,* Z-26.
40. *Water,* 184.
41. *Ibid.,* 124.
42. Water Resources Council, "Synthetic Fuels Development for the Upper Colorado Region Water Assessment," 46 Fed. Reg. 35054-35070, 6 July 1981.
43. Riparian rights prevail except for modifications in some states. For instance, Pennsylvania requires new major water consumers to provide their own storage to replace consumptive use during dry periods. Indiana does not allow new water withdrawals from navigable streams that would reduce streamflow below the existing 7-day/10-year low flow; Ohio River Basin Commission, *Synfuels in the Ohio River Basin: A Water Resources Assessment of Emerging Coal Technologies*, Cincinnati, Ohio, 1980, 49.
44. John Harte and Mohamed El-Gasseir, "Energy and Water," in *Science* 199:4329, 10 February 1978, 626.
45. Kentucky Department for Natural Resources and Environmental Protection, Russell Barnett, *Environmental Assessment of a Coal Conversion Industry in Western Kentucky*, Frankrort, KY., 1980, 53.
46. Laura King, "Ecological Consequences of the Cooling Requirements of Electrical Power Generation," in John Harte, *et al., Energy and the Fate of Ecosystems* ,National Academy Press, Washington, D.C., 1980, 346, 349.
47. U.S. Environmental Protection Agency, *Energy from the West, Policy Analysis Report* ,Washington, D.C., EPA 600/7-79-083, 102.
48. Water Resources Council, "Synthetic Fuels Development for the Upper Colorado Region Water Assessment," 46 CFR 35054, 6 July 1981.
49. Office of Technology Assessment, *An Assessment of Oil Shale Technologies*, Washington, D.C.: Office of Technology Assessment, 1980, 378.
50. The Supreme Court first enunciated the principle of reserved water rights for Indian lands in *U.S.* v. *Winters,* 207 U-S. 564 (1908). Federal rights to waters for reserved Federal lands have been upheld in a series of important cases ranging from *Federal Power Commission* v. *Oregon* 349 U.S. 435 (1955) generally known as the Pelton case, through *U.S.* v. *New Mexico* 438 U.S. 696.
51. U.S. Council on Environmental Quality, *Desertification of the United States,* U.S. Government Printing Office, 1981, 66.
52. "ETSI Passes Another Milestone in Its Travels and Travails," *The Energy Daily* 10:16, 26 January 1982, 4.

53. *Mitre,* 117.
54. *Lurgi PCGD,* 3-23.
55. *Lurgi PCGD,* 3-26, 27.
56. 40 CFR 60, Subpart Ka.
57. *National Program,* 152.
58. *SRC-II,* C-58.
59. *Lurgi PCGD,* 3-20.
60. University of Oklahoma, *Environmental Issues of Synthetic Transportation Fuels from Coal—Draft Background Report* [hereafter "Oklahoma"], Norman, OK.: Science and Public Policy Program, March 5, 1981, 111-36.
61. *Synfuels,* 3-23.
62. *1980 Keystone Coal Industry Manual* (New York, N.Y.: McGraw-Hill, Inc., 1980), p. 687.
63. *National Program,* p. 210.
64. *The Energy Daily* (March 1, 1982), p. 1.
65. *SRC-I* vol. I, pp. 4-9, 4-10.
66. Althouse, *et al.,* "An Evaluation of Chemicals and Industrial Processes Associated with Cancer in Humans Based on Human and Animal Data: IARC Monographs 1 to 20," in *Cancer Research* vol. 40, pp. 1-12 (1980) (hereafter "Cancer").
67. E. Sawicki, *et al.,* "Quantitative Composition of the Urban Atmosphere in Terms of Polynuclear Heterocyclic Compounds and Aliphatic and Polynuclear Aromatic Hydrocarbons," in *International Journal of Air and Water Pollution* vol. 9, pp. 515-524.
68. *Synfuels,* p. 3-11.
69. R.J. Sexton, C.S. Weil, and N.I. Condra, *Archives of Environmental Health,* vol. I, pp. 181-231 (1960).
70. National Institute for Occupational Safety and Health, *Occupational Exposure to Coke Oven Emissions* (Washington, D.C.: U.S. GPO, HSM 73-11016, 1973); National Institute for Occupational Safety and Health, *Occupational Hazard Assessment: Coal Liquefaction* (Washington, D.C.: U.S. GPO, 1981); H.W. Lloyd, "Long-Term Mortality Study of Steelworkers v. Respiratory Cancer in Coke Plant Workers," in *Journal of Occupational Medicine* vol. 13, pp. 53-65 (1971); C.K. Redmond, *et al.,* "Long-term Mortality Study of Steel Workers," in *Journal of Occupational Medicine* vol. 14, pp. 621-629 (1971).
71. Though the Sasol plant in South Africa has been in operation for 24 years, in-depth records of employees' (or former employees') health have not been kept.
72. E. Bingham and W. Barkley, "Bioassay of Complex Mixtures Derived from Fossil Fuels" in *Environmental Health Perspectives* vol. 80, pp. 157-164 (1979).
73. *SRC-I* vol. I, p. 4-3.
74. McAnn and Ames, "A Simple Method for Detecting Environmental Carcinogens as Mutagens," 271 *Annals N.Y. Academy of Sciences* 5 (1976).
75. Batelle-Pacific Northwest Laboratory, *Biomedical Studies on Solvent Refined Coal (SRC-II) Liquefaction Materials: A Status Report* (Richland, WA.: Batelle Laboratory, 1979).
76. Memo from James M. Evans, *et al.,* Enviro Control, Inc., to Murray L. Cohen, July 11, 1977, p. 3.
77. *Recommended Health and Safety Guidelines for Coal Gasification Pilot Plants,* (Rockville, MD: June 1980), p. 40.

78. *Environmental Review of Synthetic Fuels,* (hereafter EPA/Industrial Energy Research Laboratory, June 1980), p. 9.
79. John W. Sheehy, *Control Technology for Worker Exposure to Coke Oven Emissions* (Cincinnati, OH.: DHHS (NIOSH) No. 81-114, March 1980), p. 4.
80. *Environmental Review,* March 1980, p. 7.
81. Richard Brown, ed., *Health and Environmental Effects of Coal Tehnologies: Research Needs* (Mclean, VA.: The Mitre Corporation, 1980), pp. 51-52.
82. *Impacts,* p. 80.
83. Ronald J. Yang, *et al.,* "Coal Gasification and Occupational Health," *American Industrial Hygiene Association Journal,* vol. 39, no. 12 (December 1978), p. 986.
84. *Ibid.,* 990.
85. National Institute for Occupational Safety and Health, *Occupational Hazard Assessment: Coal Liquefaction* vol. I-Summary (Washington, D.C.: U.S. GPO DHHS (NIOSH) No. 81-131, 1981), p. 4.
86. See generally the sources cited in Norman L. Dean, Jr. *Comments on the Draft Environmental Impact Statement for the Proposed SRC-I Demonstration Plant* (Washington, D.C.: National Wildlife Federation, 1981), p. 48.
87. *Impacts,* p. 81.
88. In contrast, the families of workers may be exposed to high concentrations of carcinogens carried home on the skin and clothes of workers.
89. T.J. Mason and F.W. McKay, *U.S. County by County Mortality Study 1950-1969,* DHHS Pub. No. NIH-74-615 (1973).
90. W.J. Blot, L.A. Briton, W. Fraumeni, and B.T. Stone, "Cancer Mortality in U.S. Counties with Petroleum Industries," in *Science* vol. 198, pp. 51-53 (1977).
91. State of California, Department of Health Services, *Lung Cancer in Contra Costa County:* 1969-1979, Resource for Cancer Epidemiology (21 October 1981).
92. National Institute of Occupational Safety and Health, *Coal Gasification and Industrial Hygiene Characteristics: A Literature Review to March 1979* draft (hereafter "Ind. Hygiene Characs.") (Morgantown, W.V.: DHHS (NIOSH), March 1979), p. 1-28.
93. David T. Deutsch, "No Health Risks are Seen Yet for Synfuel Workers" in *Chemical Engineering* 9 February 1981, pp. 51-53.
94. Karl J. Bombaugh, *et al.,* "An Environmentally Based Evaluation of the Multimedia Discharges from the Kosovo Lurgi Coal Gasification System," Radian Corporation, Austin, Texas, revised April, 1981, pp. 1, 6 (hereafter "Kosovo Study").
95. *Kosovo Study,* p. 1.
96. *Kosovo Study,* p. 31.
97. *Kosovo Study,* pp. 1, 28.
98. *Kosovo Study,* pp. 1, 32, 34, 36.
99. *Kosovo Study,* p. 34.
100. *Kosovo Study,* p. 47.
101. Ronald K. Patterson, "Ambient Air Downwind of the Kosovo Gasification Complex: A Compendium," for presentation at the Symposium on Environmental Aspects of Fuel Conversion Technology-V, St. Louis, Missouri, September 16-19, 1980, p. 3, (hereafter "Kosovo Air").
102. *Kosovo Air,* p. 9.

103. *Ibid.*
104. *Kosovo Air,* pp. 9-10.
105. Environmental Protection Agency, *Project Summary: Aerosol Characterization of Ambient Air Near a Commercial Lurgi Coal Gasification Plant,* (Research Triangle Park, N.C.: U.S. EPA, EPA-600/S7-80-177, April 1981), pp. 5-6.
106. *Kosovo Air,* p. 10.
107. *National Institute for Occupational Safety and Health, Trip Report,* Ruhrgas AG, December 13, 1977, p. 8.
108. *Ibid.,* p. 6.
109. *Id.,* p. 14.
110. U.S. Army Corps of Engineers, *Final Environmental Impact Statement— Tennessee Valley Authority Coal Gasification Project* (hereafter "TVA") (Norris, TN.: TVA, 1981), pp. 2-34, 2-38.
111. *TVA,* p. 2-38.
112. NIOSH trip report, Modderfontein No. 4 Ammonia Plant, December 7, 1977, p. 3, *et seq.*
113. *Ibid.,* p. 4.
114. *Ibid.,* p. 3.
115. C.A. Zee, *et al.,* "Environmental Assessment: Source Test and Evaluation Report, Koppers-Totzek Process," No. 4 Ammonia Plant, Modderfontein, RSA, prepared for U.S. EPA/Industrial Energy Research Laboratory, EPA-600/7-81-009, Janaury 1981, p. 6 (hereafter "K-T Process").
116. *K-T Process,* pp. 38, 41.
117. *K-T Process,* pp. 41.
118. *K-T Process,* pp. 37, 48.
119. *K-T Process,* pp. ii.
120. Krupp-Koppers GmbH, "Environmental Assessment of the Koppers-Totzek Process," prepared for TRW, Inc., Essen, February, 1980, included as Appendix A in TRW report, (hereafter "Krupp-Koppers").
121. *TVA,* p. 2-34.
122. Krupp-Koppers, p. 65.
123. Krupp-Koppers, p. 71.
124. *TVA,* p. 2-33.
125. *TVA,* pp. 2-37, 38.

5. Environmental Impacts of Oil Shale Development

1. U.S. Department of Energy, *Synthetic Fuels and the Environment: An Environmental and Regulatory Impacts Analysis,* DOE/EV-0087, Washington, DC, June 1980 , [hereafter "Synthetic Fuels"], 5-78.

2. U.S. Environmental Protection Agency, *Environmental Perspective on the Emerging Oil Shale Industry* EPA-600/2-80-205a,Cincinnati, OH, 1980, [hereafter "Environmental Perspective"], p. 86.

3. *Synthetic Fuels,* 3-4.

4. *Environmental Perspective,* 85.

5. *Ibid.,* 84,86.

6. *Synthetic Fuels,* 3-4.

7. U.S. Department of the Interior, Bureau of Land Management, *Draft Environmental Impact Statement, Proposed Development of Oil Shale Resources,* (U.S. Department of Interior, DES-75-62, 1975), IV-104.

8. U.S. Office of Technology Assessment, *An Assessment of Oil Shale Technologies*, Washington, DC, June 1980, [hereafter OTA], two volumes, I-332.

9. *Ibid.,* 334-5.

10. *Ibid.,* 340.

11. Richard Brown, ed., The Mitre Corporation, *Health Effects of Oil Shale Technology,* DOE/HEW/EPA-02, MTR-79W00136, May 1979), [hereafter "Mitre"], 181.

12. R.F. Probstein and H. Gold, *Water in Synthetic Fuel Production,* Cambridge, MIT Press, 1978, 204.

13. *Environmental Perspective,* 99.

14. *Ibid.*

15. *Ibid.*

16. *Synthetic Fuels,* 3-3.

17. *Environmental Perspective,* 94.

18. *Ibid.,* 73.

19. G. C. Slawson, Jr., ed., *Groundwater Quality Monitoring of Western Oil Shale Development,* EPA-600/7-79-023, Las Vegas, NV, January 1979, [hereafter "Groundwater"], 53.

20. *Ibid.,* 58.

21. *OTA,* 308.

22. *Ibid.*

23. *Ibid.,* 309.

24. *Ibid.,* 309-10.

25. *OTA,* 385.

26. *Ibid.,* 386.

27. *Ibid.*

28. *Ibid.,* 387.

29. *Ibid.,* 256.

30. *Ibid.,* 316.

31. *Ibid.*, 317-18.
32. *Ibid.*
33. *Ibid.*, 319.
34. R. P. White, "Modern Views on Some Aspects of the Occupational Dermatoses," *Journal of Industrial Hygiene,* (Vol. 8, 1926), 367-81.
35. A. Scott, "The Occupation Dematoses of the paraffin Workers of the Scottish Oil Shale Industry," *British Medical Journal,* (Vol. 2, 1922), 381-5.
36. M. Purdue and S. Etlin, "Cancer Patterns in the Oil Shale Area of the Estonian Soviet Socialist Republic," *Environmental Health Perspectives,* (Vol. 30, 1979), 209-10.
37. *OTA,* 321.
38. *MITRE,,* 87.
39. *Ibid.,* 78-81.

6. Socioeconomic Impacts

1. U.S. Department of Energy, *Synthetic Fuels and the Environment: An Environmental and Regulatory Impacts Analysis* [hereafter "Synfuels"] DOE/EV-0087, Washington DC, June 1980, 5-122.
2. *Ibid.*, 5-157.
3. National Academy of Sciences, Committee on Surface Mining and Reclamation, *Surface Mining of Non-Coal Minerals* [hereafter "COSMAR Study"] Washington, DC, 1979, 297.
4. U.S. Energy Research and Development Administration, *Final Environmental Impact Statement: Alternative Fuels Demonstration Program* [hereafter "Alternative Fuels"] ERDA-1547, Washington, DC, September 1977, IV-87.
5. Alternative Fuels, IV-51, IV-87, IV-120, IV-154.
6. Synfuels, 5-147.
7. Synfuels, 5-137.
8. Synfuels, 5-136.
9. Synfuels, 5-138.
11. *Ibid.*
12. Synfuels, 5-143.
13. U.S. Department of Energy, *Socioeconomic Impact Assessment: A Methodology Applied to Synthetic Fuels,* [hereafter "Socioeconomic Impact Assessment"], HCP/L2516-01, 1978, 73.
14. U.S. Comission on Civil Rights, *Energy Resource Development: Implications for Women and Minorities in the Intermountain West,* Washington, DC, US GPO, 1978, [hereafter "Women and Minorities"], 18.

15. "Colorado shale boom would generate revenue to cover community impacts," *Synfuels,* 10 October 1980, 4.

16. U.S. Department of Housing and Urban Development, *Rapid Growth from Energy Projects: Ideas for State and Local Action: A Program Guide* (Washington, DC US GPO, 1976) [hereafter "HUD Report"] 30.

17. Denver Research Institute, "Analysis of Financing Problems in Coal and Oil Shale Boom Towns "[hereafter "FEA Report"], prepared for Federal Energy Administration, Denver, CO, 1976, 61.

18. FEA Report, 63.

19. HUD Report, 19.

20. U.S. Department of the Interior, *Final Environmental Impact Statement—Green River-Hams Fork Regional Coal,* two volumes, [hereafter "Green River-Hams Fork EIS"] Washington, DC, 1980, I-91.

21. HUD Report, 19.

22. "Garfield threatens Utah permits," *Synfuels Week,* 4 May 1981, 1.

23. J.S. Gilmore and M.K. Duff, *Boom Town Growth Management: A Case Study of Rock Springs—Green River, Wyoming* (Boulder, CO: Westview Press, 1975), 2.

24. U.S. General Accounting Office, *U.S. Coal Development—Promises, Uncertainties,* [hereafter "GAO Report"] 1977, 7.5.

25. HUD Report, 26.

26. "Wheatland strives for boomtown perfection," *High Country News,* 1 June 1979, 4.

27. "Wheatland strives for boomtown perfection," *High Country News,* 1 June 1979, 4.

28. Women and Minorities, 34, 45.

29. R.L. Little, "Some social consequences of boomtowns," 53 *North Dakota Law Review* 401, 1977 [hereafter "Social Consequences"], 414.

30. Socioeconomic Impact Assessment, 64.

31. Green River-Hams Fork EIS, I- 220, II-10-14.

32. J.S. Gilmore, "Boom or Bust: Energy Development in the Western United States" [hereafter "Gilmore"],in R.J. Burby and A.F. Bell, eds., *Energy and the Community* (Massachusetts: Ballinger Pub. Co. 1978), 104.

33. Green River-Hams Fork EIS, I- 220.

34. "The boom reaches across the tracks," *High Country News,* 16 November 1979, 5.

35. "Colorado plans to aid senior citizens hurt by shale's boomtown effects," *Synfuels,* 20 March 1981, 8.

36. "Union Oil settling school, worker questions, *Synfuels Week,* 11 May 1981, 4-5.

37. HUD Report, 6.

38. *Ibid.*

39. Socioeconomic Impact Assessment, 64.

40. Green River-Hams Fork EIS, II-10-27.

41. "Wheatland strives for boom town perfection," *High Country News,* 1 June 1979, 4.

42. D. Longhahe and J. Geyler, "Commercial development in small isolated energy impacted communities," *The Scial Science Journal,* 16 February 1979, 51-62.

43. U.S. Department of the Interior, *Final Environmental Impact Study: West-Central North Dakota Regional Environmental Impact Study on Energy Development*, 1978, 147.

44. Gilmore, 2.

45. Women and Minorities, 36.

46. Women and Minorities, 47.

47. Women and Minorities, 415.

48. Social Consequences, 410.

49. Green River-Hams Fork EIS, 112.

50. *Ibid.*

51. "Boomtown Women," *Mine Talk*, May-June 1981, 9.

52. Women and Minorities, 34, 45.

53. COSMAR Study, 26.

54. "Oil shale risks shared by nation's water users, taxpayers," *High Country News*, 17 April 1981, 7.

55. GAO Report, 735.

56. R.W. Burchell and D. Listokin, *The Fiscal Impact Handbook: Estimating Local Costs and Revenues of Land Development*, (New Brunswick, NJ: Center for Policy Research, 1978).

57. U.S. Department of Energy, *Final Environmental Impact Statement: Solvent Refined Coal-II Demonstration Project* DOE/EIS-0069, two volumes, Washington, DC, January 1981, I-4-53 to 59.

58. Green River-Hams Fork EIS, vol. II, Comment No. 33.

59. Socioeconomic Impact Assessment, 74.

60. U.S. Environmental Protection Agency, *Energy from the West Draft Policy Analysis Report*, Washington, DC, 1978, 480.

61. D. Masselli, N. Dean, with L. Ackerman, *The Impacts of Synthetic Fuels Development*, Washington, DC, National Wildlife Federation, 1981, [hereafter "Impacts"], 99.

62. Impacts, 99.

63. HUD Report, 30.

64. Office of Management and Budget, "Second Update to the 1980 Catalog of Federal Domestic Assistance," Washington, DC, GPO, 1980.

65. Impacts, 99.

66. Telephone conversation, L. King with Tim Schultz, Rio Blanco County Board of Commissioners, 22 September 1981.

67. Telephone conversation, L. King with Rick Moore, Director, Wyoming Industrial Siting Council, 22 September 1981.

68. *Ibid.*

69. U.S. Department of Energy, *Final Environmental Impact Statement Memphis Light, Gas and Water Division Industrial Fuel Gas Demonstration Project* [hereafter "Memphis EIS"], DOE/EIS-0071, Washington, DC, May 1981, 1-29.

70. Synfuels, 5-157.

71. Memphis EIS, 1-61.

FIGURES

Figure 1
U.S. Oil Shale Projects

Caitlin Shale Products	1917-1930
8-100 TPD retorts operated	1919-1920
1000 TPD retorts operated	1921-1924
Nevada-Texas-Utah (N-T-U) 6. (Batch retort burned from top down) (similar to 19th Century low-Btu coal gas units)	
40 TPD retort, Santa Maria, CA	1921-1925
U.S. Bureau of Mines	
N-T-U Tech Pilot Retort, Rifle, CO	1925-1929
Mobil Oil	
Pilot Plant, Paulboro, NJ	1943-1945
USBM	
Two-40 TPD N-T-U plants, Anvil Points, CO (produced total of about 20,000 barrels) Constructed pursuant to Synthetic Liquid Fuels Act	1947-1951
6 TPD gas combustion continuous process plant	1950-1955
150 TPD gas combustion continuous process plant	1952-1955
25 TPD gas combustion continuous process plant (Produced 11,000 bbls total before Congress cut off funding.)	1953-1955
These plants were again operated by an industry consortium under an agreement with DOI.	1964-1967
Union Oil	
350 TPD Union "A" continuous process plant, near Grand Valley, CO, total production about 20,000 bbls	1957-1959
Development Enginers, Inc./Paraho Oil Shale Project	
17 company consortium managed in part by staff from USBM operation, leased Anvil Points to develop Paraho Process. Produced over 110,000 bbls, use-tested by DOE and DOD.	1973-1978
Colony Development Co.	
1,000 TPD "Semiworks" using TOSCO II process, Grand Valley, CO. Produced total of 180,000 bbls.	1965-1972
Occidental Petroleum	
Experiments with underground "Modified-in-situ" retorting, largely successful. Logan Walsh, CO.	1972-present

Source: OTA-137-153.

Figure 2
Paraho Direct Heating Oil Shale Retorting Process

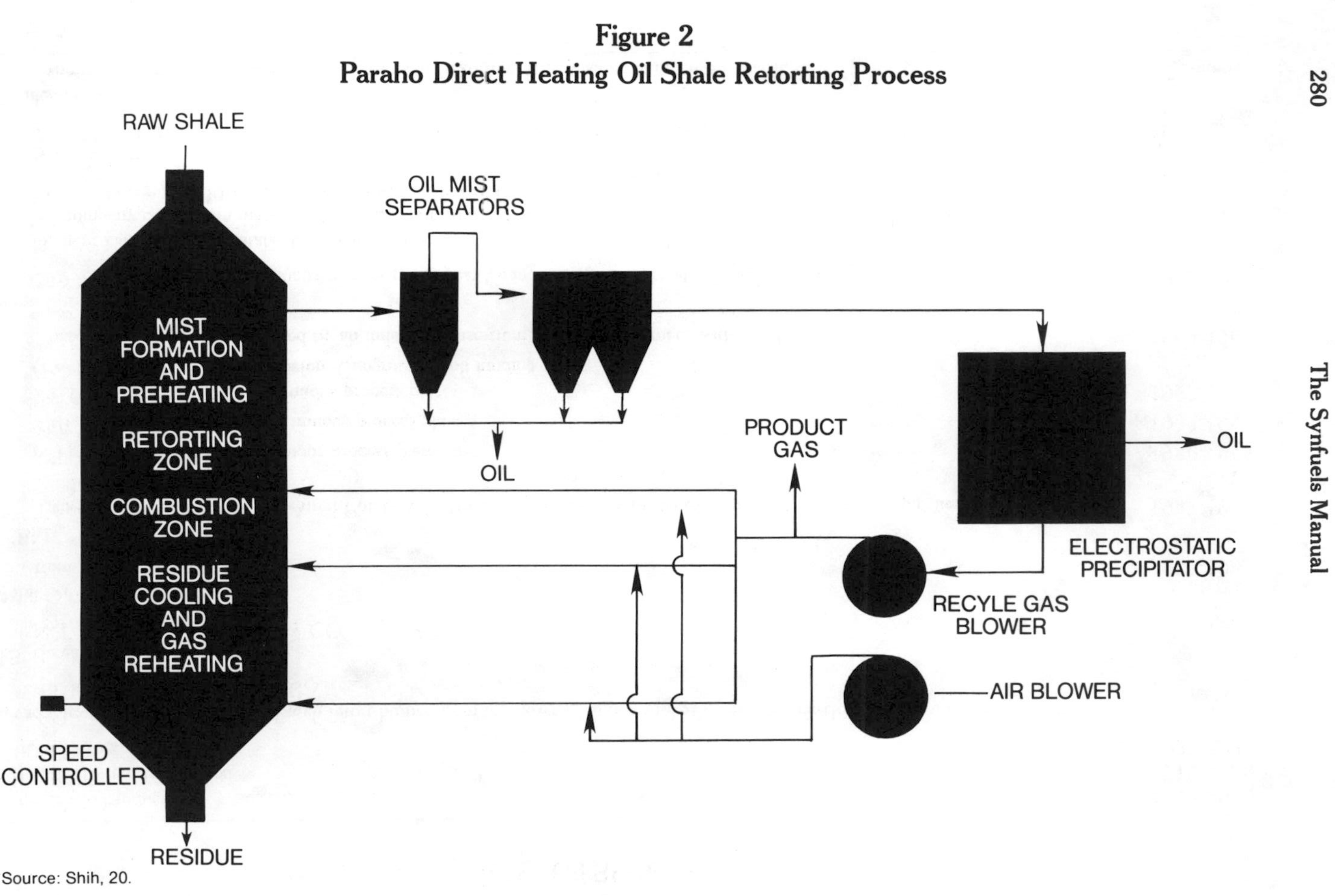

Source: Shih, 20.

Figure 3
TOSCO II Indirect Heating Oil Shale Process

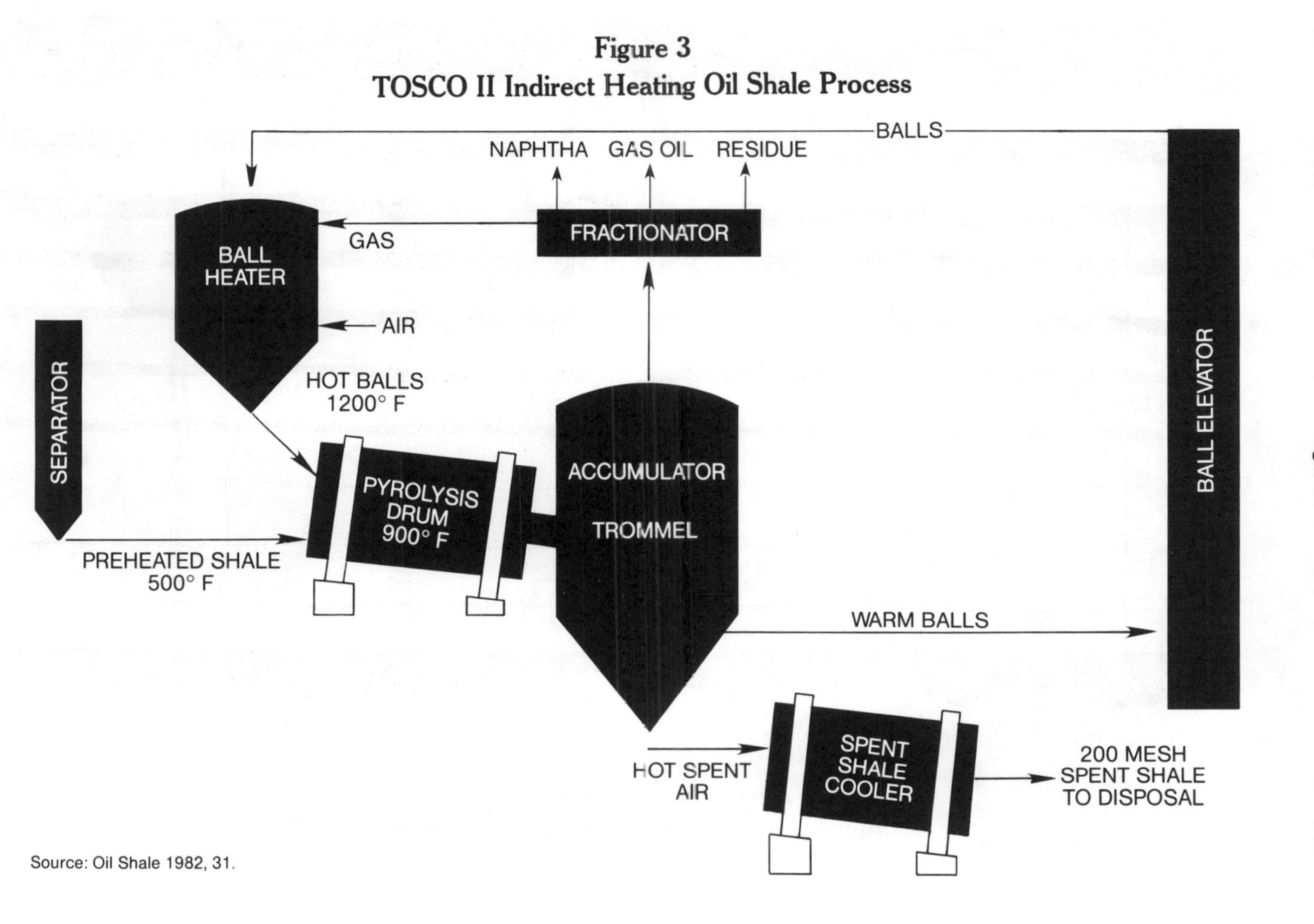

Source: Oil Shale 1982, 31.

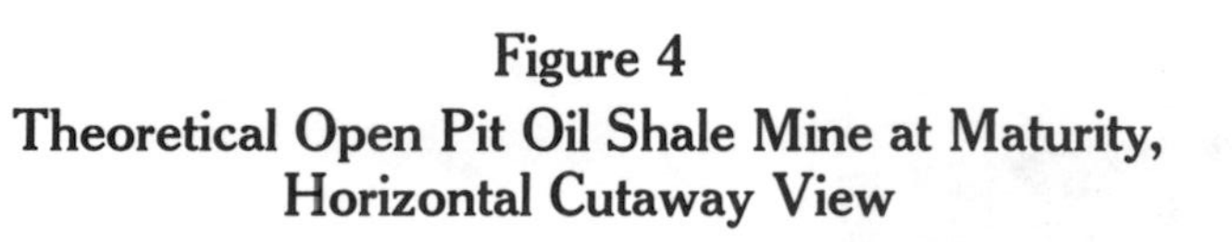

Figure 4
Theoretical Open Pit Oil Shale Mine at Maturity,
Horizontal Cutaway View

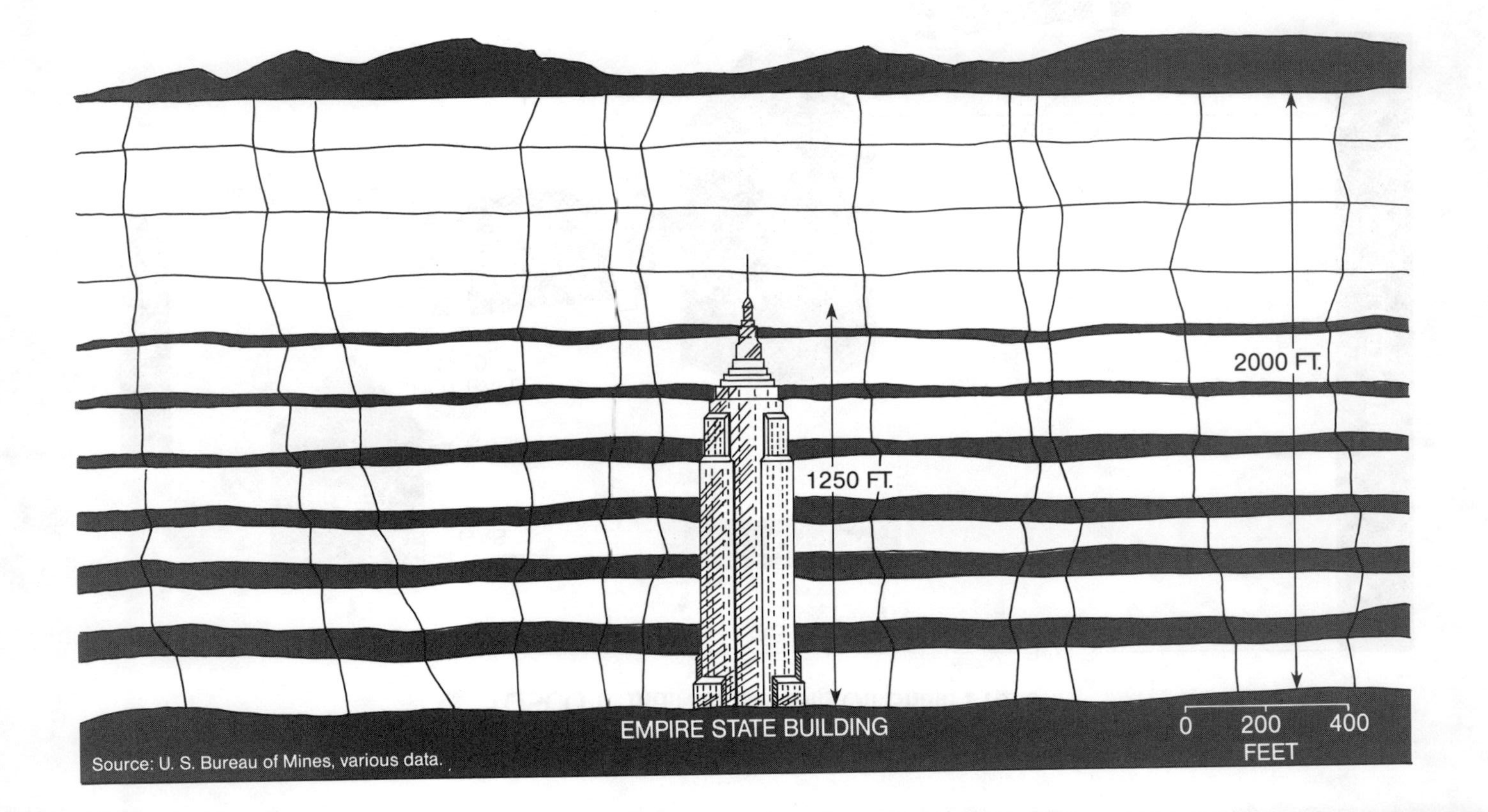

Source: U. S. Bureau of Mines, various data.

Figure 5
Theoretical Open Pit Oil Shale Mine at Maturity, Overhead View

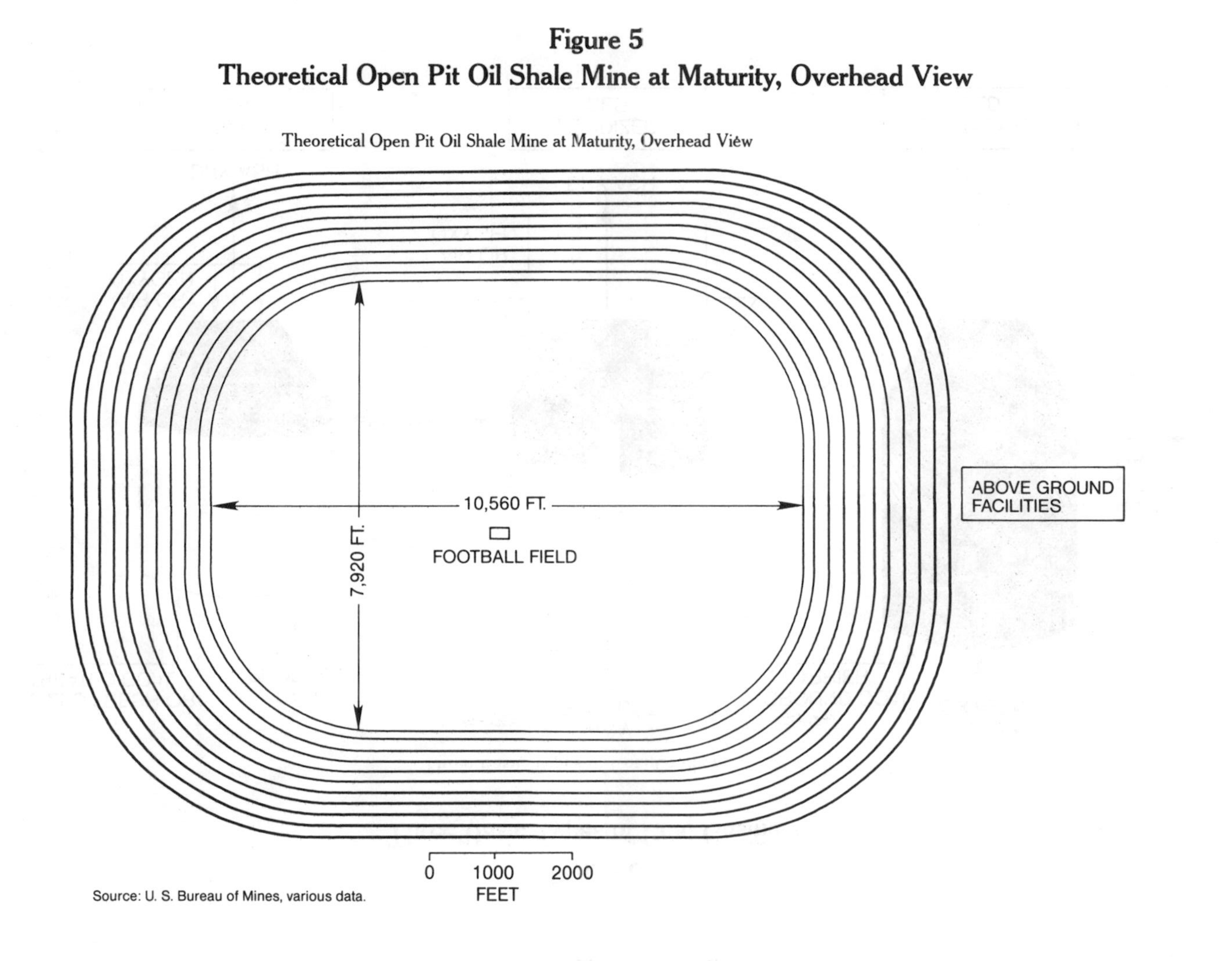

Source: U. S. Bureau of Mines, various data.

Figure 6
Three Basic Types of Coal Gasifiers

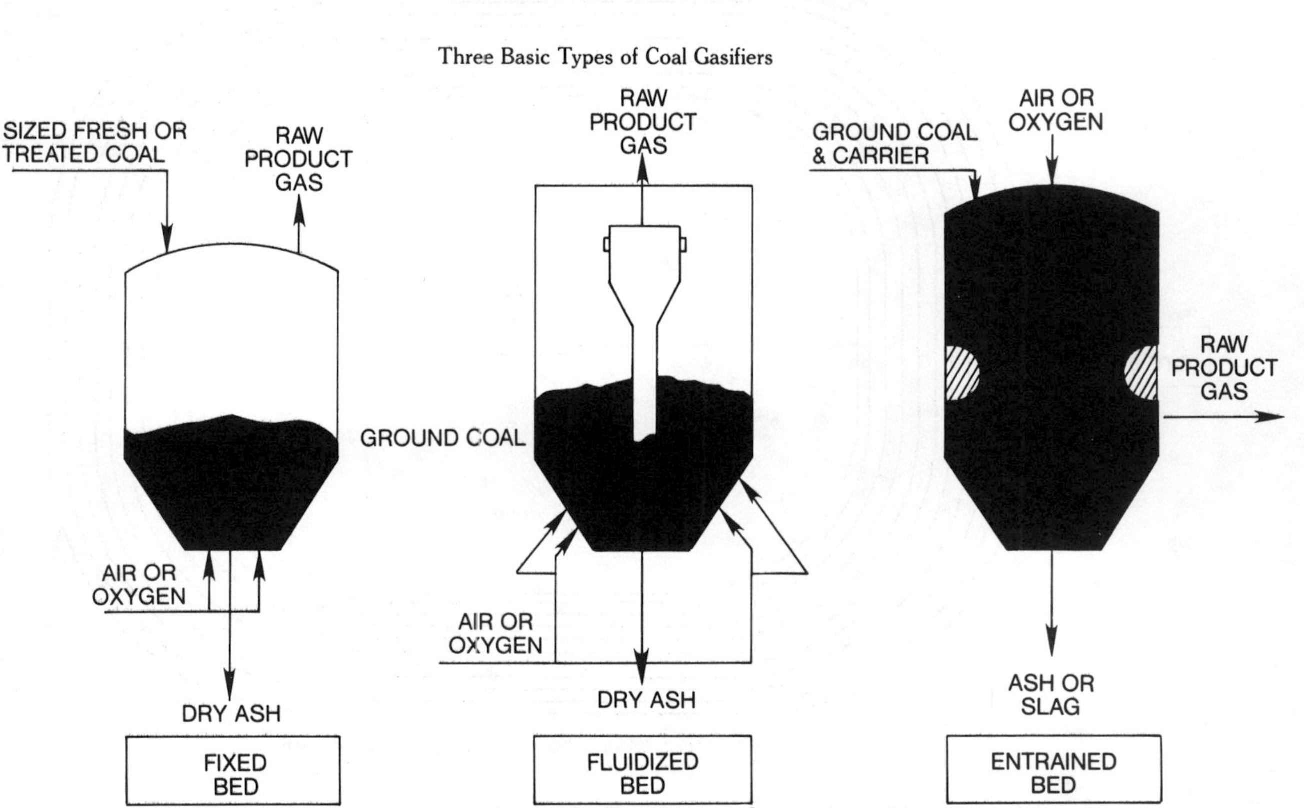

Figure 7
Lurgi Gasifier

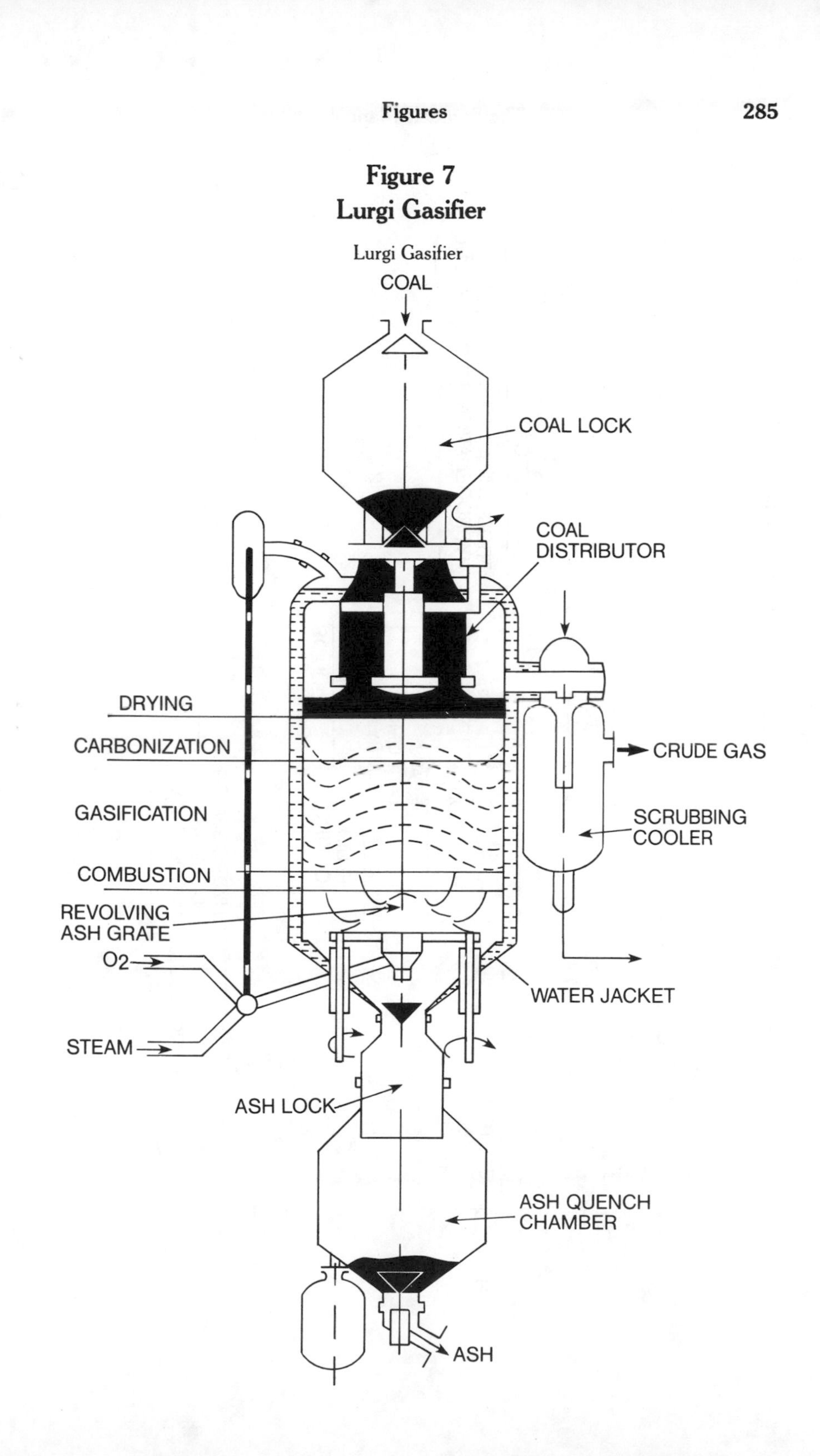

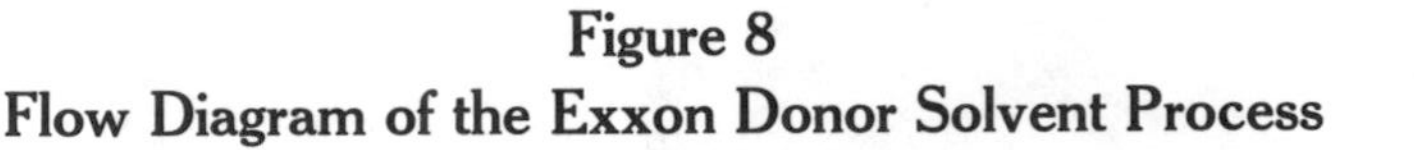

Figure 8
Flow Diagram of the Exxon Donor Solvent Process

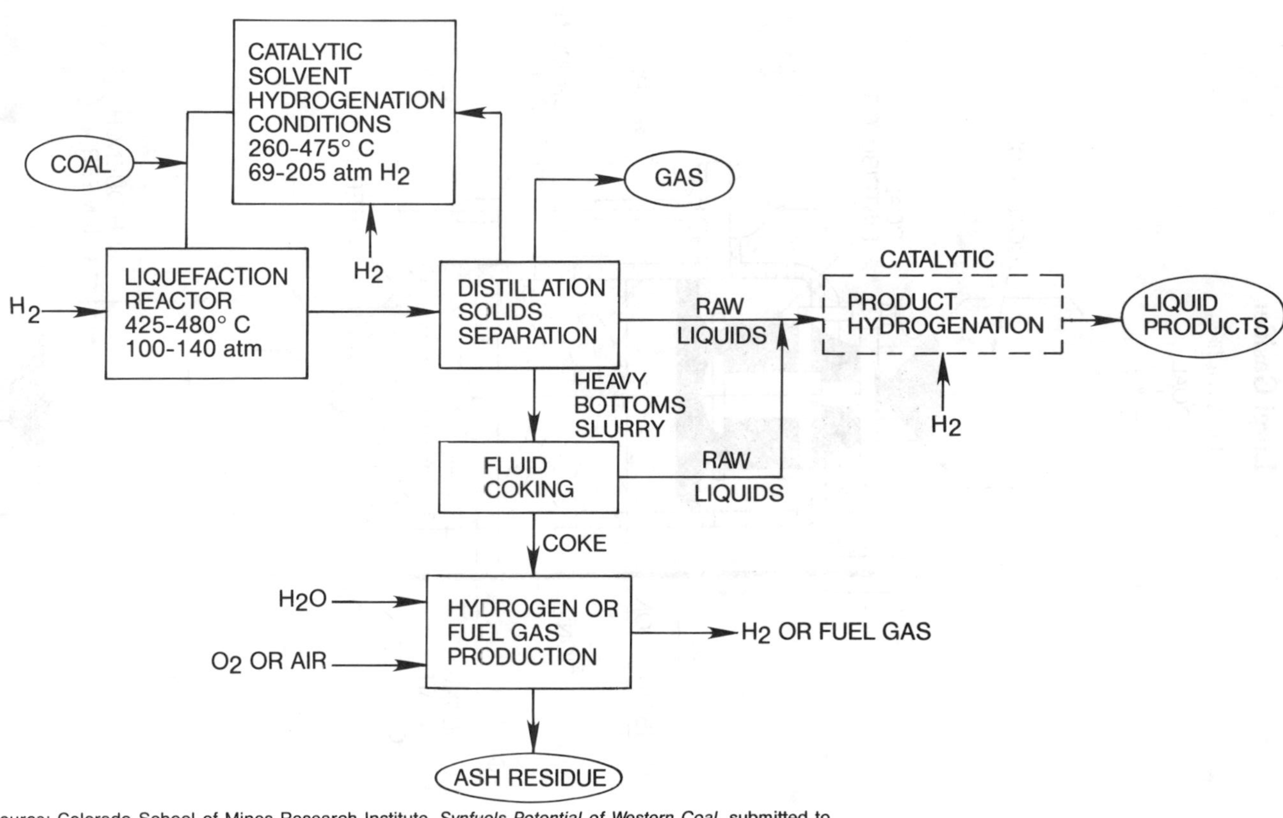

Source: Colorado School of Mines Research Institute, *Synfuels Potential of Western Coal,* submitted to Office of Technology Assessment, October 1980, 79.

Figure 9
Direct Liquefaction of Coal

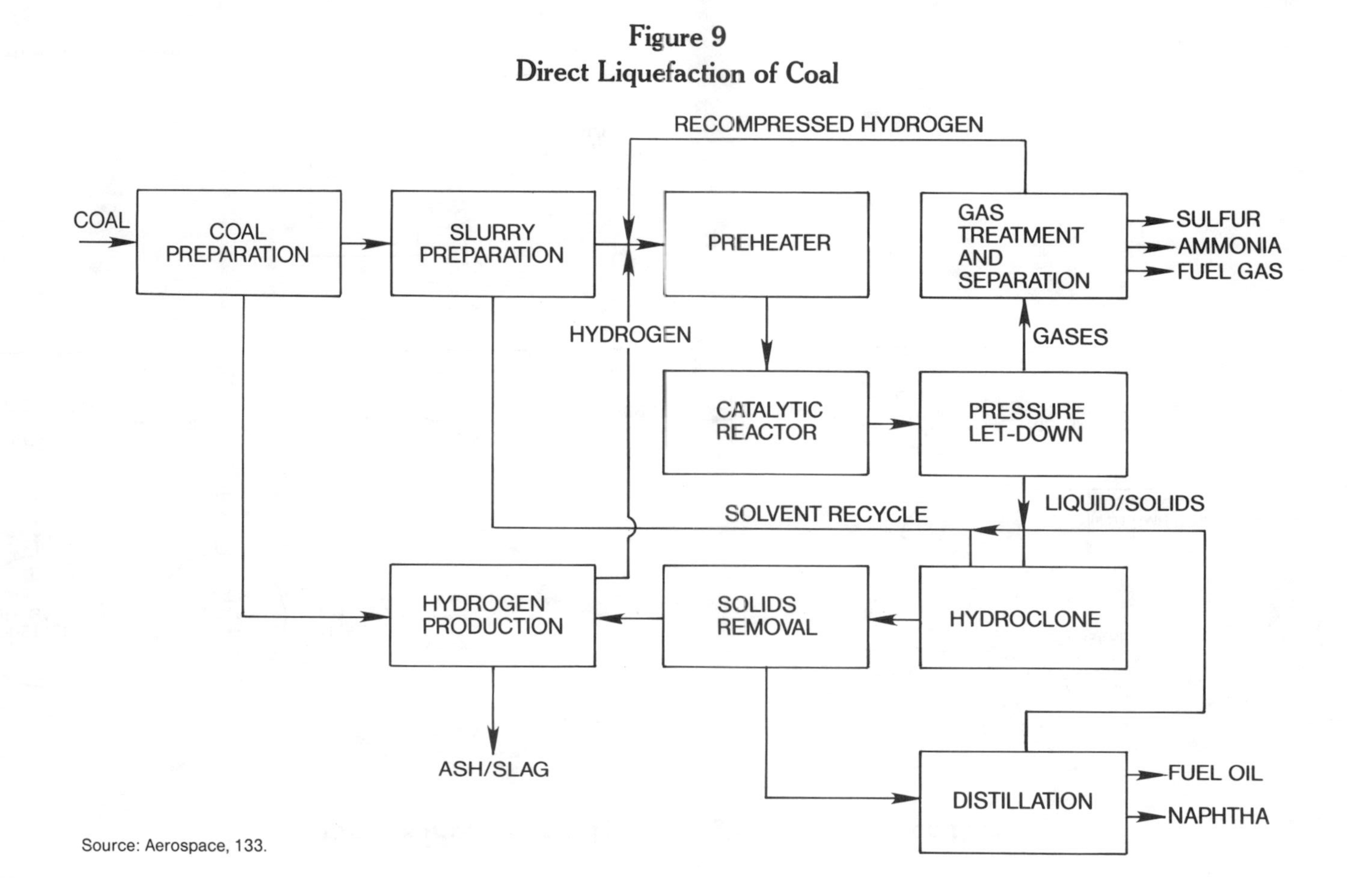

Source: Aerospace, 133.

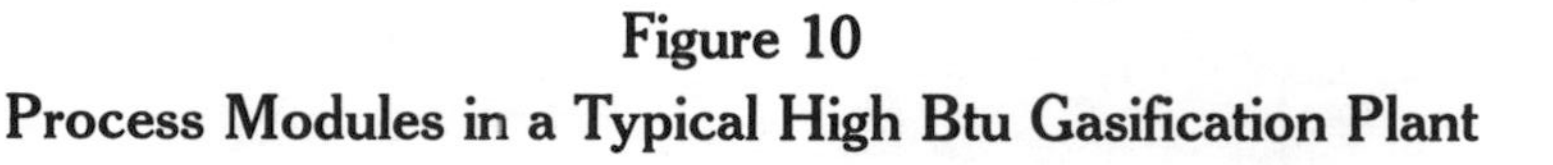

Figure 10
Process Modules in a Typical High Btu Gasification Plant

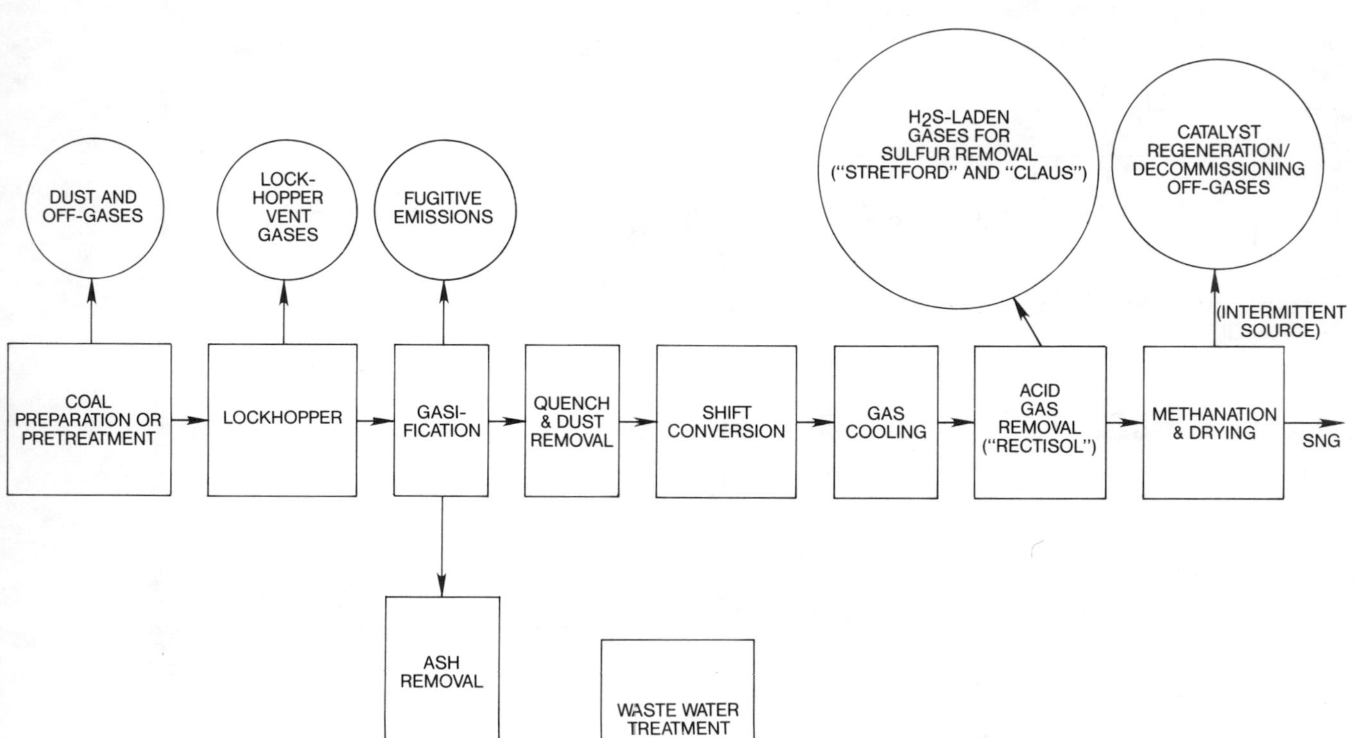

Adapted from: Gasification Data Base, I-65

Figure 11

Process Requirements and Pollutants Associated With Synfuel Conversion Technologies, Based on Preliminary Design Studies (Normalized to 50,000 Barrels Per Day Crude Oil Equivalent)

Process	Input (tons/day)	Conversion Thermal Efficiency (%)	Water* Requirements (acre-ft/yr)	Air Emissions with Controls (tons/day)					CO_2 (tons/day)	Solid Waste (tons/day)
				SO_x	NO_x	CO	HC	TSP		
Oil Shale (Surface)	125,000 (25 gal/ton) shale)	65-70	6,000-10,000	1-5	10-30	1-5	1-5	3-10	18,000	Spent shale - 108,000 Shale dust and coke - 4,000
Oil Shale (Modified In Situ)	140,000 (25 gal/ton) shale)	58-63	2,000-4,000	1-3	5-20	1-2	1-2	1-2	20,000	Raw shale** - 45,000
Fischer-Tropsch (Indirect Liquefaction)	31,000 (Subbituminous 8,500 Btu/lb) .45% sulfur	56-60	11,000-12,000	9-14	5-8	NA	NA	1-1.5	21,000	Coal prep refuse (Neglibile thick seam) Ash - 1,800-2,000 Sulfur - 65-70
Exxon Donor Solvent (Direct Liquefaction)	20,000 (Bituminous 4% sulfur	60-64	7,500-8,500	16-18	5-6	1-1.5	.01-1	.3-4	14,000	Coal prep refuse - 8,000-9,000 Ash - 2,500-3,000 Sulfur - 500-600
Solvent Refined Coal (SRC) (Direct Liquefaction)	21,000 (Bituminous 3.5% sulfur	58-62	5,500-6,500	5-10	9-10	1.5-2.5	1-2	2-3	14,000	Coal prep refuse - 8,000-9,000 Ash and slag - 5,000-7,000 Sulfur - 450-500
Mobil Technology Coal-to-Methanol-to-Gasoline	31,000 (Subbituminous 8,500 Btu/lb) .45% sulfur	56-68	11,000-12,000	9-14	5-8	NA	NA	.5-1.5	21,000	Coal prep refuse (Negligible thick seam) Ash - 1,800-2,100 Sulfur - 65-70
Lurgi Dry Ash (High-Btu Coal Gasification, 250 MMSCF/day)	38,000 (Lignite 6.783 But/lb) .6% sulfur	58-60	8,000-10,000	35-40	16-17	NA	NA	2-3	29,000	Coal prep refuse (Negligible thick seam) Ash (coker) - 2,500-3,000 Sulfur - 500-600
Coal-Fired Powerplant*** (1330 MMe)	12,750 (Bituminous 12,000 But/lb 2% sulfur	36 (direct firing)	15,000-20,000	76	35	— (Negligible)	—	5	31,000	Coal prep refuse - 3,000-4,000 Ash - 1,500 Sludge - 1,600 100% solids)

*The actual water requirements for a specific site can be substantially reduced through maximum use of dry cooling (Ref. 5) and water contained in raw coal feed.

**Can be surface retorted.

***The conventional coal-fired plant assumes state-of-the-art lime/limestone nonregenerable SO_x scrubbing, electrostatic precipitators, combustion modification NO_x control, and wet cooling towers as a basis for the estimates. Current NSPS for utility boilers was also assumed.

Note: No attempt has been made to show emissions from the end-use applications of synfuels, e.g., fuel for boilers, motor fuels.

Source: *Synfuels*, 3-23.

Figure 12
National Ambient Air Quality Standards (NAAQS)

Pollutant	Maximum Allowed Concentrations ug/m^3 (except CO)	Averaging Period
Carbon monoxide (CO)	40 mg/m^3 [a]	1-hour
	10 mg/m^3 [a]	8-hour
Hydrocarbons (HC)	235[a,b,c]	3-hour
Lead (Pb)	1.5[a,b]	3-month
Ozone (O_3)	235[a]	1-hour
Nitrogen Dioxide (NO_2)	100[a,b]	Annual
Sulfur Dioxide (SO_2)	80[a]	Annual
	365[a]	24-hour
	1300[b]	3-hour
Total Suspended Particulates TSP)	75[a]	Annual
	260[a]	24-hour
	60[e]	Annual
	150[a]	24-hour

[a]Indicates a primary standard.
[b]Indicates a secondary standard.
[c]The HC standard is a quide to be met to assist in attaining the O_3 standard.
[d]A single excursion per year (projected or measured) is permitted for all but then annual standards.
[e]Indicates a guideline for the short-term secondary standard.

Figure 13
Controlled Emissions of Air Pollutants from Gasification and Liquefaction Plants
(pounds/hour)

Process	Emissions*				
	SO_x	NO_x	Particulates	Hydrocarbons	CO
Lurgi Dry Ash Gasification	2917-3333	1333-1417	167-250	NA	NA
Fischer-Tropsch Indirect-Liquefaction	750-1167	417-667	83-125	NA	NA
Mobil M Liquefaction	750-1167	417-667	42-125	NA	NA
SRCII Liquefaction	417-833	750-833	167-250	83-166	125-208
EDS Liquefaction	1333-1500	417-500	25-33	1-83	83-125
Coal-Fired Power Plant**	6333	2917	417	—Negligible—	

*Normalized for a 50,000 b/d facility
**1330 MW

Source: DOE, *Synthetic Fuels and the Environment*, 3-23.

Figure 14
Air Pollutants Generated by a 50,000 Barrel/Day Surface Retort*
(pounds per hour)

	Pollutants Generated	Pollutants Emitted After Controls
Particulates	30,220	240
SO_2	32,390	280
NO_x	1,910	1,930**
Hydrocarbons	700	360
CO	510	500

Source: OTA, I-262, 278.

**These estimates apparently were taken from a Colony PSD application. No explanation is provided for the increase in NO_x after controls. In reality, a decrease would be expected.

Figure 15
Comparison of Air Pollution from Oil Shale Conversion with Coal-Fired Power Plant
(pounds per hour)

	Surface Retort (TOSCO II)*	Modified In Situ Retorts*	Coal-Fired Power Plant**
Particulates	240	220	11,350
SO_2	280	270	3,400
NO_x	1,930	3,500	680
Hydrocarbons	360	120	—
CO	500	210	—

*Source: OTA, I-278-279. Assumes a 50,000 bbl/day operation.
**U.S. Bureau of Land Management, *Final Environmental Impact Statement on Allen-Warner Valley Energy System*, Vol. 1, p. 4-12 (December 1980). Figures are projected emissions from the 2,000 MW Harry Allen coal plant, using Best Available Control Technology.

Figure 16
Volume of Solid Wastes from Oil Shale Production*
(tons per day)

Type of Waste	Volume
Spent oil shale	100,000
Oil upgrading catalysts	1,625
Raw oil shale	240
Spent filters	136
Water treatment plant sludges	3 (dry; 300 wet)
Miscellaneous landfill (garbage, etc.)	3 (dry; 300 wet)
Sewage sludge	2.5 (wet; 0.5 dry)

*For a 50,000 b/d facility.

Source: G.C. Slawson, General Electric Co.—TEMPO, *Groundwater Quality Monitoring of Western Oil Shale Development*, [hereafter "Groundwater"] prepared for EPA, EPA-600/7-79-023, Las Vegas, NV, January 1979, 193-4.

Figure 17
Pollutants Generated by Surface and Modified In Situ Oil Shale Processing

			Emissions (pounds per hour)		
Stage	Particulates (Surface[b]/MIS[c])	SO_2 (Surface/MIS)	NO_x (Surface/MIS)	Hydro-carbons (Surface/MIS)	CO (Surface/MIS)
Mining	1,480/4,540	0/0	250/300	50/120	440/180
Shale preparation	15,940/—	0/—	0/—	0/—	0/—
Retorting	11,440/0	150/0	1,430/0	480/0	60/0
Spent shale treatment & disposal	1,350/450	0/0	130/100	10/10	0/0
Upgrading	trace/10	10/10	20/80	10/trace	trace/10
Ammonia & Sulfur recovery	0/0	32,000/24,000	0/0	0/0	0/0
Product storage	0/0	0/0	0/0	150/80	0/0
Steam & power	—/20	—/trace	—/2,800	—/0	—/0
Hydrogen production	10/80	30/20	80/220	trace/20	10/20
TOTAL	30,220/5,100	32,399/24,030	1,910/3,500	700/230	510/210

[a]Emissions from a 50,000 b/d facility.
[b]All surface retort emissions estimated for TOSCO II process, using room and pillar mining.
[c]All MIS retort emissions estimated assuming underground mining.

Source: OTA, I-262-3

Figure 18
Cooling and Process Water Requirements of Oil Shale Technologies
(gallons per million Btu fuel)

	Cooling Water	Process Water	Sub-total
Paraho Direct	11.6	(+ 1.16)*	10.4
Paraho Indirect	13.3	0.20	13.5
TOSCO II	8.5	2.81	11.3

Adapted from Probstein and Gold, 124, 185, 27.

Note: Cooling Water estimates assume the use of a technology of medium efficiency.

*This process produces more water than it uses.

Figure 19
Summary of Uncontrolled Gaseous Waste Streams Generated in Lurgi-Based Indirect Liquefaction Facilities

Stream	Pollutants of Concern	Factors Affecting Stream Flow and Characteristics	Uncontrolled Waste Stream Flow[a] ($Nm^3/10^9kcal$ of coal to gasifier)
Waste Streams Main Process Train			
Coal storage/preparation			
• Raw coal storage	Particulates	Characteristics of coal storage pile, meteorological conditions	N/A
• Prepared coal storage, feeding and pelletizing	Particulates	System design and quantity of coal processed.	N/A
• Crushing, screening and transfer	Particulates	System design, quantity of coal crushed/ screened size requirements for gasifier	N/A
High pressure coal lockhopper vent gases	Reduced sulfur compounds, particulates, RHC, CO, POM, HCN, NH_3	Characteristics of pressurant gas, cycle frequency, coal rank and sulfur content	15,000 - 18,000
Low pressure coal lockhopper vent gases	Reduced sulfur compounds, particulates, RHC, CO, POM, HCN, NH_3	Characteristics of pressurant gas, cycle frequency, depressurization cut-off pressure	1,700 - 2,000
Ash lockhopper vent gases	Particulates, HCN, H_2S, NH_3, RHC	Coal ash content, coal rank	1,600 - 2,400
Transient waste gases	Reduced sulfur compounds, RHC, CO, particulates, HCN, NH_3, POM	Gasifier down time/operational difficulties, coal rank and sulfur content	9,900 - 19,000 [520 - 1000][b]
H_2S-rich acid gas	Reduced sulfur compounds, RHC, CO, HCN, NH_3	Type of AGR process used (selective or non-selective), coal rank and sulfur content	29,700 - 102,000
CO_2-rich acid gas	Reduced sulfur compounds, RHC, CO, HCN, NH_3	Type of AGR process used (selective or non-selective), coal rank and sulfur content	0 - 68,500
Catalyst regeneration/ decommissioning off-gases			
• Shift	Reduced sulfur compounds, SO_2, CO, particulates, trace elements	Decommissioning/regeneration frequency and duration, catalyst characteristics	26,000 - 31,000 [2100 - 2500][b]
• Methanol and Methanation	Particulates, CO NI(CO)$_4$	Decommissioning/regeneration frequency and duration, catalyst characteristics	1,000 - 1,200 [3 - 4][b]
• Mobil M	Particulates, CO	Decommissioning/regeneration frequency and	1,500 - 1,800 [710 - 850][b]
Off-gases from F-T light ends recovery section	RHC, CO	Quantity of hydrocarbons and CO produced in gasification process	7600 - 7900
Mobil fractionator off-gases	RHC, CO	Quantity of hydrocarbons and CO produced in gasification process	600 - 610
Methanation condensate depressurization off-gases	RHC, CO	Flow rate of crude SNG	unknown
CO_2 vent gases from SNG production	CO	Quantity of hydrocarbons and CO produced in gasification process	610 - 1600
Fugitive Organic Emissions	RHC, other gases	Vapor pressures of liquids, temperatures and pressures of process lines, types of seals or gaskets used, maintenance practices	N/A
Auxiliary Processes			
Boiler flue gases	Particulates, NO_x, SO_2	Quantity and mix of coal, by-products and waste streams burned in the boiler; fuel characteristics, overall plant energy requirements	170,000 - 350,000
Product and by-product storage	RHC	Meterological conditions, types of tanks used, loadout procedures, vapor pressure of liquids.	N/A
Cooling tower evaporation/drift	NH_3, HCN, H_2S, RHC	Characteristics of make-up water, cooling load of facility	N/A
Pollution Control Processes			
Ammonia stripper overhead gases	Sulfur compounds, RHC, HCN, NH_3	Coal rank and sulfur content	6,200 - 7,800
Brine concentration off-gases	H_2S, NH_3, HCN, RHC	Type of upstream wastewater treatment employed, coal rank	Unknown
Fugitives from ash quench	Particulates	Coal ash content	N/A
Flue gas from wastewater incinerator	Particulates, SO_2, NO_x	Type of upstream wastewater treatment, coal rank, fuel used	
Activated carbon regeneration off-gases	Particulates, RHC, CO	Type of upstream wastewater treatment, coal rank, fuel used	200 - 2,000 [10 - 100][b]

[a]Ranges correspond to different coals and synthesis processes. N/A indicates not applicable.

[b]For intermittent streams brakets indicate annual average emission rates.

Figure 20
Net Water Consumption of Synfuel Technologies

	Gallons/million Btu fuel	Acre-feet per day at standard plant
Gasification		
Synthane	14-18	10.4-12.8
Hygas	15-19	11.0-13.8
Lurgi	18-26	13.5-18.7
Liquefaction	12-16	11.0-15.6
Solvent Refined Coal	5-11	5.2-10.7

Adapted from Ronald F. Probstein and Harris Gold, *Water in Synthetic Fuel Production: The Technology and Alternatives* [hereafter "Water"], (Cambridge: MIT Press, 1978), 268, 270.

Note: These estimates assume the use of a cooling technology of medium efficiency.

Figure 21
Cooling and Process Water Requirements of Synfuel Technologies
(gallons per million Btu fuel)

	Cooling Water	Process Water	Sub-Total
Gasification			
Synthane	6.9-8.0	5.17	12.1-13.2
Hygas	6.0-6.1	6.43	12.4-12.5
Lurgi	14.2-15.9	6.63	20.8-22.5
Liquefaction	7.9-9.1	1.55-4.83	9.45-13.9
Solvent Refined Coal	3.2-3.3	0.31	3.5-3.6

Adapted from *Water*, 124, 185, 271.

Note: Cooling water estimates assume the use of a technology of medium efficiency.

*This process produces more water than it uses.

Figure 22
Maximum and Minimum Expected Concentrations in Waste Water Streams for White River Shale Project

Stream	TDS (mg/1)		Oil and grease (mg/1)		Phenol (mg/1)		Ammonia (mg/1)		pH (mg/1)	
	Max	Min	Max	Min	Max	Min	Max	Min	Max	Min
Sour water bottoms	—	—	100	50	150	80	50	25	9.5	8.5
Oily water	2,000	500	1,000	50	500	50			9	7
Sanitary waste water	1,000	800	50	20	1	1	15	10	8.5	7
High TDS	10,000	5,000	—	—	—	—	—	—	8	7
Freshwater	800	400	—	—	—	—	1	1	8.5	7
Process water (in shale)	15,000	2,000	3,000	2,000	390	115	4,000	1,700	8.7	8.1
Weighted mean	7,253	2,225	880	506	124	44	989	420	8.3	7.6

Figure 23
Net Water Consumption of Oil Shale Technologies

	Gallons/million Btu Fuel	Acre-Reef per day at standard plant
Paraho Direct	19	16.8
Paraho Indirect	31	27.0
Tosco II	31	27.0

Adapted from Ronald F. Probstein and Harris Gold, *Water in Synthetic Fuel Production: The Technology and Alternatives* (Cambridge: MIT Press, 1978), 268, 270.

Note: These estimates assume the use of a colling technology of medium efficiency.

Figure 24
Aqueous Discharges from Coal Liquefaction Systems For Conceptualized 7,950 m^3 (50 kbbl) per day plants

	Process H-Coal				EDS			
	Raw wastes		After treatment**		Raw wastes		After treatment**	
	Metric tons/day	(tons/day)	Metric tons/day	(tons/day)	Metric tons/day	(tons/day)	Metric tons/day	(tons/day)
Sour water (process)								
Total (including water)	4,557	(4,913)	4,154	(4,580)	1,974	(2,176)	1,776	(1,958)
H_2S	175	(193)		0.15ppm		N.A.		0.15ppm
Oil								
Phenols	23.6	(26)		0.1 ppm	18	(20)		0.1 ppm
NH_3	103	(114)		10 ppm	180	(198)		10 ppm
Hydrogen generation****								
Clarifier overflow	419	(462)	419	(462)	172	(190)		N.A.
Other wastewater	1,204	(1,327)	1,204	(1,327)	841	(927)		N.A.
Cooling tower blowdown								
Total	563	(621)	563	(621)		N.A.		N.A.
Coal preparation								
Total	*	*	------	------	11,000**	(12,127)**		N.A.
Water	------	------	------	------	7,109	(7,837)		N.A.
Tailings	------	------	------	------	3,828	(4,220)		N.A.
Coal pile runoff***	63.5	(70)	63.5	(70)	63.5	(70)	63.5	(70)

Until thorough environmental sampling and analyses of all streams have been performed, these lists will not be complete.

*Using dry coal preparation.
**Using water technique for extensive coal preparation.
***Rainfall estimate (based on Illinois rainfall).
****Koppers-Totzek process

N.A.—Not available

Source: *EPA Liquefaction Data Base*, II-xxv.

Figure 25
Foul Water Analysis, Rocky Flats Pilot Run (TOSCO PROCESS)

Analysis	Concentration (mg/ l)[a]			
	Production day			
	1	2	3	4
Inorganics				
Total dissolved solids (organics removed)	6,660	1,980	5,940	15,300
ph	8.6	8.7	8.1	8.6
Specific conductance (mhos per cm)	12,500	14,800	15,300	13,300
Calcium (Ca)	45	9.8	25	12
Magnesium (Mg)	<0.1	6.2	<0.1	19
Sodium (Na)	<1.0	<1.0	<1.0	<1.0
Potassium (K)	<5.0	<5.0	<5.0	<5.0
Arsenic (As)	0.07	0.09	0.08	0.06
Selenium (Se)	0.03	0.04	0.03	0.05
Molybdenum (Mo)	<1.0	<1.0	<1.0	<1.0
Lithium (Li)	<10	<10	<10	<10
Bicarbonate (HCO_3)	5,400	12,600	6,920	12,900
Carbonate (CO_3)	1,560	2,550	2,130	2,850
Chloride (C l)	1,300	855	1,090	1,160
Flouride (F)	<1.0	<1.0	<1.0	<1.0
Cyanide (Cn)	<0.01	<0.01	0.01	<0.01
Silica	8	4	12	12
Nitrate (NO_3)	330	330	320	170
Phosphate (PO_4)	0.5	0.21	15.6	0.5
Ammonia (NH_3)	3,685	4,025	3,960	1,740
Total sulfur	855	1,210	775	1,240
Sulfide	848	1,200	768	1,230
Sulfate	8	8	8	8
Elemental sulfur	<1.0	<1.0	<1.0	<1.0
Organics				
Neutral oil	2,560	2,840	2,115	1,950
Amines	602	856	916	1,600
Carboxylic acids	6,480	1,680	1,215	515
Phenols	390	220	270	115
Total organic carbon	16,300	21,000	18,200	14,200
Chemical oxygen demand (COD)	24,600	31,000	27,100	23,500
5-day biochemical oxygen demand (BOD_5)	10,800	9,400	27,100	9,000

[a]Units are mg/ l unless otherwise indicated. Values of pH are standard units.

Figure 26
Potential Synfuel Constituents and Their Health Effects

Compound	Major Health Effects
Aldehydes and Ketones	
Formaldehyde	Cancer, Respiratory irritation, Skin and eye irritation
Acrolein	Respiratory irritation, Skin and eye irritation
Aromatic Amines	
Aniline	Cancer, Methemoglobinemia (chemical asphpyxiant)
4 Aminobiphenyl	Cancer
Combustion Gases	
Sulfur oxides	Respiratory irritation, Skin and eye irritation
Nitrogen oxides	Respiratory irritation, Skin and eye irritation
Carbon monoxide	Asphyxiation
Cyanides	
Hydrogen	Asphyxiation, Skin irritation
Ammonium thiocyanate	Asphyxiation
Hydrocarbons	
Benzene	Cancer, Chromosomal damage, Aplastic anemia, Central nervous system (CNS) impairment
Toluene	Skin and eye irritation
Organometallics	
Nickel Carbonyl	Cancer, Respiratory irritation and fibrosis
Particulates	
Coal dust	Respiratory fibrosis and irritation
Phenols	
Phenol	Cancer (promoter), Skin and eye irritant, Chromosomal damage
Polycyclic Aromatic Hydrocarbons (PAHS's)	
Benzo[a]pyrene	Cancer
Dibenz[a,h]anthracene	Cancer
Benz[a]anthracene	Cancer
Radionuclides	
Radon	Cancer
Radium	Cancer
Sulfur Compounds	
Hydrogen Sulfide	Birth defects, CNS impairment
Carbon disulfide	Birth defects, Reproductive and developmental effects
Thiophene	CNS impairment
Mercaptans	CNS impairment
Trace Elements	
Arsenic	Cancer
Beryllium	Cancer
Cadmium	Cancer, Chromosomal damage, Birth defects, Reproductive and developmental effects
Chromium	Cancer, Respiratory irritation and allergic reaction
Lead	Cancer, Reproductive effects, CNS impairment, Kidney damage, Anemia
Mercury	Birth defects, CNS impairment, Kidney damage
Nickel	Cancer, Dermatitis (allergic reaction)

Figure 27
Proportion of Low-Skill Employment In Synfuel Plant Construction (Man Hours)

Plant Type	Size	Source/Date		Low Skill Employment Rate	Average Number
High BTU Gasification	250 $\overline{M}$scf/d	ERDA	1977	13.0%	128
High BTU Gasification	250 $\overline{M}$scf/d	EPA	1979	15.0%	281
High BTU Gasification	250 $\overline{M}$scf/d	IGT	1981	12.3%	214
Medium BTU Gasification	2,500 $\overline{M}$scf/d	ERDA	1977	12.0%	181
Low BTU Gasification	2,500 $\overline{M}$scf/d	IGT	1980	11.4%	54
Low BTU Gasification	10 $\overline{M}$scf/d	IGT	1980	12.2%	51
Fischer-Tropsch Liquefaction	100,000 bpd	EPA	1979	18.6%	357
Non-Fischer Tropsch Liquefaction	50,000 bpd	ERDA	1977	8.0%	153
SRC-II Liquefaction	20,000 bpd	DOE	1981	11.9%	174

Source: *Argonne Urban Facilities*, 18.

Figure 28
Division of Responsibilities under Clean Air Act (PSD) and Clean Water Act (NPDES)

	NPDES (as of) 22 June 82)	PSD (as of 31 March 82)
Region 1		
Connecticut	x°	complete, await EPA
Maine		SIP
Massachusetts		complete, conditions
New Hampshire		complete, conditions
Rhode Island		complete, conditions
Vermont	x°	SIP
Region 2		
New Jersey	x°±	SIP
New York	x ±	complete, OMB review
Region 3		
Delaware	x	SIP
D.C.		complete, await EPA
Maryland	x	complete, await EPA
Pennsylvania	x ±	incomplete
Virginia	x ±	complete, await EPA
West Virginia	x°±	complete, conditions
Region 4		
Alabama	x°±	SIP
Florida		SIP
Georgia	x°±	SIP
Kentucky		incomplete
Mississippi	x°	SIP
North Carolina	x°	SIP
South Carolina	x°±	SIP
Tennessee	x	complete, await EPA
Region 5		
Illinois	x ±	incomplete
Indiana	x ±	incomplete
Michigan	x ±	complete, conditions
Minnesota	x°±	complete, await EPA
Ohio	x	complete, conditions
Wisconsin	x°±	incomplete
Region 6		
Arkansas		SIP
Louisiana		SIP
New Mexico		complete, conditions
Oklahoma		SIP
Texas		complete, await EPA
Region 7		
Iowa	x°±	incomplete
Kansas	x	incomplete
Missouri	x°±	incomplete
Nebraska	x ±	incomplete
Region 8		
Colorado	x	complete, conditions
Montana	x ±	complete, await EPA
North Dakota	x	SIP
South Dakota		SIP
Utah		incomplete
Wyoming	x ±	SIP
Region 9		
Arizona		incomplete
California	x ±	incomplete
Hawaii	x ±	incomplete
Nevada	x ±	complete, await EPA
Region 10		
Alaska		incomplete
Idaho		complete, await EPA
Oregon	x°±	incomplete
Washington	x	complete, await EPA

x = state has been delegated NPDES program
° = state also has approved pretreatment program
± = state also approved to regulate federal facilities under NPDES
SIP = state has no outstanding Part D SIP conditions or actions
complete = state has submitted a SIP judged by EPA to be complete, but not yet approved fully.
conditions = EPA has made final approval subject to certain conditions which the sate must fulfill and has not yet.
await EPA = state has fulfilled conditions and EPA has not yet given final approval.
OMB review = approved by EPA but undergoing review by OMB.
incomplete = state submittal not judged complete by EPA, or state has not submitted SIP.

Sources: CAA/PSD information: "Status of Part D SIP's as of March 31, 1982," memo from Q.T. Helms, Chief, Control Programs Operations Branch, to Chief, Air Programs Branch, Regions I-X; U.S. EPA, Research Triangle Park, NC, 23 April 1982.

CWA/NPDES information: untitled document dated 22 June 1982 from U.S. EPA, Washington, D.C.